C0 1 72 71473 31

SHIP "BETSEY," CAPTAIN FANNING, RETURNING TO PORT
From a lithograph in Fanning's *Voyages*, New York, 1833

Voyages & Discoveries in the South Seas 1792–1832

EDMUND FANNING

DOVER PUBLICATIONS, INC., NEW YORK

PUBLISHER'S NOTE, 1989

The present Dover edition does not alter the text of the 1924 edition, which was very conservative with regard to the forms and spellings of many of Fanning's terms. Thus the reader should not be jolted by the presence of more than one form of the same geographical name (e.g., "Fee Jee" alongside the currently accepted "Fiji"). In some cases neither or none of the printed forms is correct ("Cape de Verds" and "Cape de Verde(s)" for the Cape Verde Islands). Polynesian names are naturally garbled. Aside from some island names that are no longer in use, the only old Yankee form that might really cause difficulty in identification is "Mas(s)afuero" for Más Afuera. Accents are generally missing, as in the cases of Curaçao and Juan Fernández. Names of people and other terms suffer, too: "Bouganville" for Bougainville, "albercore" for albacore, "sheerwater" for shearwater, "Brittanic" for Britannic, etc. Punctuation and other features are likely to be equally erratic. The reader is urged to consider this a historical text rather than a modern gazetteer.

Published in Canada by General Publishing Company, Ltd., 30 Lesmill Road, Don Mills, Toronto, Ontario.

Published in the United Kingdom by Constable and Company, Ltd., 10 Orange Street, London WC2H 7EG.

This Dover edition, first published in 1989, is a republication of the work originally published in 1924 as Publication Society, Salem, Massachusetts. The positio and a very sketchy map of the southwest Atl omitted.

Manufactured in the United States of /
Dover Publications, Inc., 31 East 2nd

Library of Congress Cataloging-in-Publicatic

Fanning, Edmund, 1769–1841.
[Voyages round the world]
Voyages and discoveries in the South Seas, 1792–1832 / Edmund Fanning.
p.   cm.
Originally published in 1833 under title: Voyages round the world.
ISBN 0-486-25960-9
1. Fanning, Edmund, 1769–1841—Journeys.   2. Voyages around the world.   I. Title.
G440.F3   1989
910.4′1—dc19   88-32723
                                                                      CIP

INTRODUCTION TO THIS EDITION

ACCOUNTS of voyages into unknown seas and among strange peoples have always possessed a keen interest for the stay-at-home traveler and ever since the days of Richard Hakluyt, sailors' narratives of adventure and discovery have filled the imagination of English readers and stimulated further ventures. It was not until after the Revolution, however, when trade with many European ports became more difficult, that American shipmasters and owners to any extent, began to search for new markets and to explore the commercial possibilities of all parts of the world. The islands of the Pacific, the west coast of North and South America, and the open ports of China and the East Indies were then visited by small sailing vessels hailing from Boston, Salem, New York and other Atlantic ports and a trade in furs, oil and Oriental products increased rapidly. Unfortunately but few of these Yankee sailors left behind them accounts of their voyages and so the narrative of Capt. Edmund Fanning of Stonington, Conn., which found its way into type in after years, must possess an unusual interest.

Fanning sailed for the South Seas in May, 1792, on a voyage for seal skins and during the next twenty-five years made voyages to the Pacific and around the world; visiting China and Australia, the desolate lands of South Georgia and the savage islanders at Fiji and the Marquesas. He discovered several hitherto unknown islands and Fanning's Island, lying twelve hundred miles south of Honolulu, still bears his name although it is now a British possession. In 1829 he was instrumental in sending out an exploring expedition to the South Seas under the command of Captains Benjamin Pendleton and Nathaniel Brown Palmer and it was his petition to Congress and largely his own personal efforts that finally led to the Act of May 14, 1836, authorizing the sending out of the exploring expedition that sailed under the command of Commodore Charles Wilkes. Following his voyage in the *Betsey*, in 1792, Captain Fanning either commanded or acted as directive agent for upwards of seventy expeditions to the South Seas and China. He died in New York City, April 23, 1841, aged 71 years.

Captain Fanning's *Voyages* was first published with the following title:—

Voyages Round the World; with selected Sketches of Voyages to the South Seas, North and South Pacific Oceans, China, etc., performed under the command and agency of the author, . . . New York, *Collins & Hannay,* 8vo., pp. 499, 5 plates, MDCCC-XXXIII.

London and Paris editions were issued the next year. The second edition, with additions, was published under a different title, viz.:—

Voyages to the South Seas, Indian and Pacific Oceans, . . . North-West Coast, Feejee Islands, . . . with an account of the New Discoveries made in the Southern Hemisphere between the Years 1830-1837 . . . Second Edition. New York: *William H. Vermilye,* 12mo., pp. 324, plate, 1838.

Three other editions followed with the same pagination as the second and all editions have now become scarce and seldom appear in the book shops.

To the general reader Fanning's *Voyages* is probably one of the most interesting of all the early voyages made by Americans that have been published and it deserves a wide acquaintance among those who would learn of the adventurous courage that led the Yankee sailor to set his course in every sea.

In the preparation of this edition of Fanning's *Voyages,* the officers at the Peabody Museum, Salem, have been most helpful. Cordial thanks are also due to Mr. Charles C. Willoughby, Director of the Peabody Museum of Archæology and Ethnology, Cambridge; Mr. John H. Edmunds, Massachusetts State Archivist; Captain G. A. Pentecost, R.N.R.; Mr. Victor H. Paltsits of the New York Public Library; Mr. A. J. Wall, Librarian of the New York Historical Society; Mr. Herbert Putnam, Librarian of Congress; the National Geographic Society and the Pan American Union, Washington, D. C.

INTRODUCTION

IN presenting to the public the following narration of Voyages, &c., the author has thought proper to introduce the same with a sketch of the earlier preriod of his life premising that the unsettled state of the country at that time, engaged in war with Great Britain, possessing but few opportunities, and those few confined mostly to the cities, in the improvement of which the inhabitants could obtain ought more than an ordinary education; together with the little leisure that a seaman's profession, ever since steadily prosecuted, has allowed him, to increase the instruction obtained from the tutor in charge of the school of his native borough, and the possibility that it would ever fall to his lot to publish any thing connected with himself never entering his mind, must plead in excuse for any grammatical errors that criticism may discover in these pages. In fact, nothing has induced him to place their contents before the public eye, but an earnest desire that his friends and countrymen may obtain such information, as while it shall please and instruct by bringing to the circle around the fire-side events as they occurred in those far distant climes, shall also lay before others whose business may call them that way, such suggestions as when faithfully followed will promote their advantage, and be of service to them.

The author's father, Gilbert Fanning, was one of four brothers, viz. Thomas, Phineas, Gilbert, and Edmund. Of these, Thomas and Edmund (from the last of whom the present Captain Edmund is named), embraced the Royal cause, and were warmly engaged in furthering the interests of his Britannic majesty. The latter, during a considerable portion of the struggle, was in command as colonel of a regiment, in the city of New York; he afterwards rose to the rank of general, and concluded his earthly career at his residence in Portman Square, London, in the year 1813.

Phineas and Gilbert, on the contrary, chose the side of the infant colonies, and throughout that eventful period, so truly described as "trying men's souls," were earnest and ardent advocates in favor of the liberties of their country. Gilbert, being in the commissary line, was actively engaged in providing and supplying the army under General Washington: he resided at Stonington, county of New London, state of Connecticut, at which place I was born, July 16th, 1769.

There were in our family eight brothers: the eldest, Nathaniel, was a midshipman in the navy, and private secretary to the celebrated Captain John Paul Jones, at the time of the bloody engagement between his ship, the Good Man Richard, and H. B. M ship of war, the Serapis. He died during the second term of Mr. Jefferson's administration, while in command at the United States Naval Station, Charleston, S. C.

The next eldest, Gilbert, was captured while on a cruise in the private armed schooner Weasel, and with Thomas, a younger brother, with him at the time, was taken to New York and confined on board the Jersey prison-ship, where Gilbert, after an unfortunate failure in an attempt to escape, died, and his bones now moulder in peace, mingled with those of other patriots who perished on board this loathsome ship. Shortly after his death, the before mentioned Colonel Edmund Fanning, then stationed in the city, being informed of Gilbert's death, and in addition to this learning that Thomas was dangerously sick, procured his removal to a lodging provided for him in the city, where, after much good nursing, he recovered, was exchanged, and returned to his parents in Connecticut.

All the other brothers followed the sea, and are all deceased; myself, by the goodness of Heaven, still remaining among the living, and tendering with much respect this plain history of voyages to my fellow-citizens, confident that at least my nautical brethren may deem them worth a perusal.

CONTENTS

I.	EARLY VOYAGES TO THE WEST INDIES .	1
II.	FIRST VOYAGE TO THE SOUTH SEAS . .	4
III.	VOYAGE TO ENGLAND	14
IV.	VOYAGES TO THE WEST INDIES . . .	33
V.	FIRST VOYAGE ROUND THE WORLD . .	40
VI.	PASSAGE TO THE CAPE DE VERDES . .	44
VII.	PATAGONIA AND THE FALKLAND ISLANDS	49
VIII.	PASSAGE ROUND CAPE HORN	65
IX.	MASSAFUERO TO MARQUESAS ISLANDS .	73
X.	MARQUESAS AND WASHINGTON ISLANDS	85
XI.	AT NUGGOHEEVA ISLAND	108
XII.	NUGGOHEEVA TO CANTON	153
XIII.	AT MACAO AND CANTON	183
XIV.	CANTON TO NEW YORK	195
XV.	AROUND THE WORLD IN THE "ASPASIA"	206
XVI.	VOYAGE OF THE BRIG "UNION" . . .	230
XVII.	VOYAGE OF THE SHIP "CATHARINE" .	242
XVIII.	SEA ELEPHANTS AND FUR SEALS . . .	255
XIX.	VOYAGE OF THE SHIP "VOLUNTEER" .	265
XX.	SUNDRY VOYAGES TO THE SOUTH SEAS	295
XXI.	MODERN ASSERTED DISCOVERIES . .	316
XXII.	SANDAL WOOD, BEACH LA MER, ETC. .	322
	INDEX	331

ILLUSTRATIONS

SHIP "BETSEY," CAPTAIN FANNING, RETURNING
TO PORT . . . . *Frontispiece*
  From a lithograph in Fanning's *Voyages*, New York, 1833.

VIEW OF STONINGTON, CONNECTICUT 1
  From a wood engraving in Barber's *Connecticut Historical Collections*, New Haven, 1837.

THE CEREMONY OF "CROSSING THE LINE" . . 6
  From an engraving in Drake's *Collection of Voyages*, London, 1770.

FUR SEALS AND SEA LIONS 7
  From lithographs in Scammon's *Marine Mammals*, San Francisco, 1874.

THE CAPTURE OF THE FRENCH FRIGATE "CLÉOPÂTRE" BY THE BRITISH FRIGATE "NYMPHE," CAPTAIN EDWARD PELLEW, ON JUNE 18, 1793   22
  From a colored aquatint by Cornelius Apostool after a drawing by Lieut. T. Yates, London, 1794.

THE BRITISH FRIGATE "BLANCHE" BLOCKADING NEW YORK HARBOR 23
  From an engraving in the *Naval Chronicle* (1814), after a drawing by Pocock.

VIEW OF NEW YORK HARBOR IN 1796 44
  From the engraving by St. Memin.

VIEW ON THE PATAGONIAN COAST 45
  From an engraving in Anson's *Voyage Round the World*, London, 1748.

ILLUSTRATIONS

VIEW OF MASAFUERO NEAR JUAN FERNANDEZ . 74
From an engraving in Anson's *Voyage Round the World*, London, 1748.

VIEW AT LA CHRISTIANA, MARQUESAS ISLANDS . 75
From an engraving in Shillibeer's *The Briton's Voyage*, Taunton, 1817.

THE MISSIONARY SHIP "DUFF," CAPT. WILSON 92
From an engraving published in London in 1805.

NUGGOHEEVA, MARQUESAS ISLANDS 93
From an engraving in Porter's *Voyage in the South Seas*, London, 1823.

PADDLE, CLUBS AND CHIEF'S STAFFS, MARQUESAS ISLANDS 126
From a photograph of the originals at the Peabody Museum, Salem.

MODEL OF A WAR CANOE, MARQUESAS ISLANDS . 127
From a photograph of a model brought to Salem before 1817 and now in the Peabody Museum, Salem.

A WARRIOR OF THE MARQUESAS ISLANDS . . . 144
From a photograph made about 1900.

WAR CONCH AND FOOD BOWL, MARQUESAS ISLANDS 145
From a photograph of the originals brought to Salem before 1821 and now in the Peabody Museum, Salem.

VIEW OF THE ISLAND OF TINIAN, LADRONE ISLANDS 170
From an engraving in Anson's *Voyage Round the World*, London, 1748.

A SHIP OF THE EAST INDIA COMPANY 171
From an etching by E. W. Cooke.

MACAO, CHINA 180
From an oil painting by a Chinese artist, at the Peabody Museum, Salem.

ILLUSTRATIONS

BOCA TIGRIS OR THE "TIGER'S MOUTH," AT THE
MOUTH OF THE RIVER LEADING TO CANTON,
CHINA 181
    From an oil painting by a Chinese artist showing the forts and the American ship "Telahoupa."

THE FACTORIES AT CANTON, CHINA 190
    From an oil painting by a Chinese artist, at the Peabody Museum, Salem.

MODEL OF A MALAY PIRATICAL PROA FROM THE
EASTERN COAST OF SUMATRA 191
    From a photograph of the original (before 1838) at the Peabody Museum, Salem.

THE BRITISH FRIGATE "CLEOPATRA," 32 GUNS . 208
    From an engraving in the *Naval Chronicle* (1805), after a drawing by Pocock.

SHIP "ASPASIA," CAPE HORN BEARING NORTH BY
EAST 209
    From a lithograph in Fanning's *Voyages*, New York, 1833.

WHAMPOA, CHINA 224
    From an oil painting by a Chinese artist, at the Peabody Museum, Salem.

THE "CENTURION," CAPTAIN ANSON, TAKING A
SPANISH GALLEON 225
    From an engraving in Anson's *Voyage Round the World*, London, 1748.

TONGATABOO, TONGA ISLANDS 236
    From an engraving in Wilkes' *U. S. Exploring Expedition*, Volume III, Philadelphia, 1845.

CAPE OF GOOD HOPE AND TABLE MOUNTAIN . . 237
    From a drawing by H. C. DeMillion.

A MALE AND FEMALE SEA ELEPHANT 256
    From an engraving in Anson's *Voyage Round the World*, London, 1748.

SEAL ROOKERY, BEAUCHENE ISLAND, FALKLAND
    ISLANDS 257
    From a lithograph in Fanning's *Voyages*, New York,
    1833.

SEALER'S ENCAMPMENT, BYERS ISLAND, FALK-
    LAND ISLANDS 296
    From a lithograph in Fanning's *Voyages*, New York,
    1833.

PALMER'S LAND AS SEEN FROM THE SOUTH SHET-
    LANDS 297
    From a lithograph in Fanning's *Voyages*, New York,
    1833.

*Voyages & Discoveries
in the South Seas
1792–1832*

VIEW OF STONINGTON, CONNECTICUT

From a wood engraving in Barber's *Connecticut Historical Collections*, New Haven, 1837

CHAPTER I

EARLY VOYAGES TO WEST INDIES

AT the early age of fourteen years the author commenced going to sea as cabin boy; and in the performance of several coasting and West India voyages out of Stonington and New London, rose through the regular grades of a seaman, second, and then first mate; and after performing in the last mentioned station three voyages to the West Indies, came to the resolution, in company with a kinsman, Mr. Asa Rossiter, to repair to, and sail from, the port of New York, as there were no prospects of a field opening for us, suitable to our then ambitious views and desires, at our native place. We therefore immediately repaired to that city, where upon our arrival, being entire strangers, we took a stroll among the shipping, to look up a voyage, which at last was obtained on board a brig belonging to the Messrs. Murry, Mumford & Bowen, commanded by Captain Miller, who, though unacquainted with either of us, yet consented to give us a voyage. But at this juncture, the report of his character came near discouraging my companion; for the captain had obtained among the seamen of the port, that of being the "hardest horse" a man could sail with, as that he had learned his trade, said they, in and by several voyages he had performed to Africa, after slaves: the house, however, owned several vessels, and was in good standing; there was, therefore a road to advancement and the prospect of steady employ, could we also muster determination, notwithstanding the captain's notorious severity, to persevere in the performance of our duties, and give satisfaction, we should eventually succeed, though for the present we suffered much. These arguments being duly weighed, and being previously pledged to ship and sail together, were conclusive: accordingly, we went on board the brig, bound on a voyage to the south side of Hispaniola, to be absent about six months, during which we had the good fortune to please our hard captain so well, that upon the arrival of the brig at Aux Cayes, both mates being discharged, Rossiter was chosen second, and myself as first mate, in their stead, in which capacities we performed duty the remainder of the voyage. Nevertheless, this man was beyond doubt, one of the most unfeeling officers that ever walked the quarter-deck. Scarce a night passed but we were on deck-duty until twelve or one o'clock, attending to his company on board, or waiting to receive him, with a lantern, at the gang-way, upon his return from visiting; so that since my elevation to the officer's berth, my rest, for the remainder of the voyage, was not over four in every twenty-four hours.

While at Aux Cayes, receiving our homeward cargo, there lay moored alongside for some weeks, a vessel belonging to Mr. Elias Nexsen, merchant, of New York, commanded by his brother, Captain Nexsen, from whom, on the return of our brig to New York, I received an invitation to engage in the employ of his brother, which was accepted; Captain Miller giving the usual recommendations, at the same time observing, that to remain with him I should have my own terms. The same proposition was made to Mr. Rossiter, who has since commanded some of the finest ships out of the port of New York, and declined.

This house was embarked in the Curacao trade, and to this port the author performed several voyages under Captain Hook.

CHAPTER II

FIRST VOYAGE TO THE SOUTH SEAS

THIS voyage came highly recommended to the mind of Mr. Nexsen, as an enterprise in every way worthy his attention, and promising to be, from the high value he was informed that the South Sea fur seal skins were held at in Canton, one of great profit, and although a new field, yet so encouraging, that in company with some friends, it was resolved upon sending out a vessel to procure a cargo suitable for the Chinese market. Accordingly the brig *Betsey* was taken up, and after being overhauled preparatory to so long a voyage, was fitted out with every necessary article that could be conceived as in any way likely to add to the comfort of her officers or men.

Great difficulty was experienced in finding officers acquainted with this new business, in our then young commercial community, to take the charge of the brig: however, after much delay, a Mr. William Whetton, who had performed a voyage out of England, in the Greenland whale and seal fishery, and was an able and experienced seaman, was shipped as second mate, (myself being the first); still was the expedition delayed for want of a captain. All things were ready for sea, crew shipped, &c., when a Captain R. Steele applied for the command, and assured the owners he was acquainted with seals of every kind, as also every thing connected with the sealing business: coming recommended too, unhappily for us, he was engaged as commander. The second day after, being in May 1792, we weighed anchor, and got the brig under full sail for sea.

While passing the Narrows, a dispute arose between the captain and the pilot, about the channel, during which so little attention was given to the vessel and her course, that she was run a ground on the west bank, it being then falling water, where she remained until the next flood, when at high tide we succeeded in heaving her afloat, with but trifling damage. We then proceeded on our voyage, and after a somewhat lengthy passage, arrived at the Cape De Verde Islands, and anchored in eleven fathoms of water, on the south side of the island of Fogo, for the purpose of obtaining wood, water, and refreshments.

At this island the water is wretched: it is filled from standing pools, which the inhabitants call springs. And in these, we found the cattle standing, seemingly little inclined to withdraw. Their continued moving about while in them, together with its brackishness, and being strongly impregnated with sulphur, makes the water totally unfit for sea supply, for after being a short time on board, it becomes extremely offensive, and cannot be used but in the most urgent cases. The beef, goats, hogs, and fowls, are tolerably good and cheap. The yams, potatoes, pumpkins, &c., were procured in abundant supply, and at very reasonable charges.

While at anchor in the road, the volcano of this island burst into full operation. Large stones were thrown to an immense height in the air, and with the burning lava pouring down on every side, accompanied by a loud roaring, as of distant thunder, was one of the most beautiful and awful sights ever witnessed. The decks of our vessel, though at the distance of three quarters of a mile from the shore, were covered nightly with the ashes and cinders.

On leaving Fogo, we passed near the island of Brava, upon which some beautiful green vallies were to be seen; then stood to the southeastward, and crossed the equator in about twenty-four degrees west longitude from London.

On crossing the line, our captain being in that turn of mind, caused all the crew to be confined below (with the exception of the officers and three favorites), and had them called up singly, to go through the ordeal which all persons on their first appearance in this part of the globe, have long been subjected to. Not an individual on board ever having crossed the equinoctial line, of course in justice all were liable to this portion of Neptune's strict laws, captain and all. "But," said he, "by my influence with his Oceanic Majesty, myself, officers, and my three favorites are excused." At this moment Neptune received an invitation to come on board; which being complied with, his said highness was quickly arrayed in the most singular and uncouth fashion. The ceremony was commenced, and speedily finished, our captain being greatly delighted with what, in his opinion, was a most jovial and brilliant affair. Not so, however, with the seamen; for during the remainder of the voyage, those who were subjected to it, were inveterate in their dis-

THE CEREMONY OF "CROSSING THE LINE"
From an engraving in Drake's *Collection of Voyages*, London, 1771

FUR SEALS (ABOVE) AND SEA LIONS (MALE AND FEMALE)
From lithographs in Scammon's *Marine Mammals*, San Francisco, 1874

like to all whom they thought instrumental in bringing it upon them.

As we proceeded toward the south, had sight of the island of Trinidad or Ascension, passed by its west end at about one league distant, and saw a fine stream of water falling down a precipice, for a supply of which we stood greatly in need, that on board being of a very bad quality. However, did not stop, but crowded sail to the southward for the Falkland Islands, where we arrived in the month of September.

On our arrival, learned that the seals were up in great numbers on some of the outer islands: we found here, to our great disappointment, that our captain, notwithstanding his declarations when he engaged with the owners, had not the least knowledge of the sealing business; in fact, he did not know the male from the female seal. Therefore, we were under an obligation to our friends and countrymen who had arrived at these islands, for the information requisite in taking the seals, and preserving their skins.

On the 7th of August last, after passing Trinidad, at about five bells P. M. (half past two), my watch on deck, the brig going at the rate of about one and a half miles per hour, saw an appearance off the quarter deck of something like a fish, keeping way with the vessel. Thinking it to be a large dolphin, took the harpoon, to which was attached a coil of whale line, and let drive at the object: I was fortunate in striking just abaft the main wing fin of a large shark, perhaps the easiest part of the whole body to be entered by such a weapon. We soon hauled him alongside, and by introducing a running bowline noose at the end of a large rope, over the flukes of his tail, and hooking the watch tackle to it, in a very short time had him flouncing on deck. He was of the kind seamen call the shovel-nose, and measured ten and a half feet in length.

Some years previous to this, on a winter passage from the West Indies to the United States, when being thrice driven off the coast to the southern edge of the gulf stream, and many days without any salt provisions, we caught a shark of about the same size, which was immediately cut up, and each man's portion given him: some being fried, was thought at that time to be the sweetest meat or fish ever tasted. Remembering this, our steward received directions to prepare some of the choicest cuts from our present prize, and fry them for breakfast, expecting the same to be bang-up; but, sadly to our discomfiture, when brought upon the table the following morning, it was found to be so intolerably strong, and withal of such bad odor, that it was impossible to impose it upon our olfactory nerves, much less think of eating any part of it. Our disappointment was very great, for even the men were made to expect a rare treat: and more speedily than the half picked bones of a bit of roast beef would have been, was this unsavory dish thrust out of the cabin.

Our want of knowledge in the sealing business was made very manifest in the outset; for shortly after the vessel was moored in the harbor, a party, consisting of myself, boatswain, and thirteen men, started out on the first seal-hunting excursion from the brig. Among the number there was an uncommonly large and stout Irishman, named Michael, commonly called Mike, who had been just landed in New York previous to his shipping: one of those good natured honest sort of fellows; possessed too, of a great abundance of that ready wit peculiar to his fellow-countrymen, so well calculated to excite laughter and keep men in good spirits. Mike was often on the passage out telling how he would slay the large bull-seals, and declaring to his shipmates, that with his *shillalah* he would soon kill tiers of them. Our lookout was posted on an elevated hill near the north end of the island, where two bays of some extent put in, each of which had a sandy beach at their head, divided nearly in the centre by a projecting rocky point. On the westermost beach (which was the farthest one from our lookout), some rods up from the margin of the water, lay about three hundred sea-lions. These being the first that had come within our view on the land, we took them to be a flock (or rookery, as was the term) of the real fur seals, after which we were searching. Upon this discovery, the men were divided into two parties, the boatswain taking six, with orders to make a circle on the upland at a proper distance, so as to keep from disturbing the lions, and to arrive at the rocky point on the farthermost side of the bay, at the end of the beach on which they lay. While the remaining few, Mike being of this number, under my own command, took station at the nearer rocky point between the two bays, under cover behind the rocks, there to wait until the boatswain and party should have arrived at, and left the farthermost point, when both were to advance in Indian file, and meeting, would thus cut off the seals from the water, and drive them back on the upland; (as we had been informed that by getting between the seals and the water, at the same time hallooing and raising a loud shout, they would turn to the upland, and could then be easily driven to a convenient place for butchering) our party being thus brought to within one hundred yards of the nearest of the flock, were presented with the view of these giant animals.

"Sir!" exclaimed Mike, as we were wondering whether our small vessel, of one hundred tons, could carry many thousand such mammoths, "Do you think these overgrown monsters are seals?" Surely they are, he was quickly answered; and as you are much the stoutest man among us, step here in advance, and lead on. However, his courage was altogether out of his reach at this most needy moment, and nothing can exceed the comical figure the man presented, as he stood wavering between fear to disobey, and fear to obey, thus exposing, as he expressed it, "his precious body to be devoured by the shutting of the jaws of such monstrous cratures: indeed, Sir, only look; by St. Patrick!" and he pointed, all pale and trembling to a large lion, which at this moment gaping, showed his rows of large ivory teeth, shook his long and shaggy mane, and concluded the exhibition with a tremendous roar.

Some of the men upon seeing this immediately exclaimed, "We are willing to follow Mike, sir." "Ah! honies," he replied, "swate Ireland!"

Although much perplexed at this moment, whether it was correct or prudent to risk the lives of the men within reach of such, to all appearance, powerful animals, I could not refrain from enjoying a hearty laugh, at the conclusion of this pithy dialogue.

The prize seemed too good a one to be lost without an effort to secure it, and so soon as the boatswain's signal of being ready was produced, both parties advanced against the common enemy, each proving the goodness of his lungs in striving to out shout his companions. This noise alarmed the lions, so that they immediately rose, and sent forth a roar that appeared to shake the very rocks on which we stood, and in turn advancing upon us in double-quick time, without any regard to our persons, knocked every man of us down with as much ease as if we had been pipe stems, and passing over our fallen bodies, marched with the utmost contempt to the water.

After the drove had passed, we arose, much gratified to find all had come off unharmed; but at this instant, a lion of middling size was espied just emerging from the high tussuck grass, ("this grass grows from bogs, the largest of which are rising twenty feet in circumference, down to a small size; called by the seamen, tussucks. They are strewed over the sides of the hills and vallies, and where the earth is the strongest and most favorable, growing in so thick a mass, and the blades of grass intwined in such strength, that one cannot pass between them, but must first cut a path through. The spears or stalks spring out of those bogs in several flat blades, encircling the pith over each fellow blade, and although the parent stalk generally is only about the size of a man's thumb, the blades in their rankest state grow to the height of eight or nine feet, so that a person walking through a tussuck valley is entirely under cover, and as much out of sight as he would be in the centre of a field of our Indian corn. When at its full height, at the time it puts forth its silky shoot from the ear, a valley of this tussuck grass, as waved by the breeze in a fine day, has, at a little distance, much the appearance of English grain when nearly ripe, with a pleasing green and tinged golden color,") against whom we concentrated our forces, and after a battle, during which he forced us down to the water's edge, succeeded in slaying him. He was in length about fourteen feet.

We felt at this moment, notwithstanding our being so easily overpowered by the drove, that we had obtained a victory, and in addition were convinced, to our entire satisfaction, that these were not fur seals.

The absence of Mike, during these past events, was not observed; for in the midst of so much confusion, as was caused by the first overthrow, and the after victory, every man paid strict obedience to nature's first law, caring only for himself; but now that it was time to return to the vessel, Mike was not to be found. Some of the men at last discovered him, busily engaged beating something with his club, by the point of rocks the party had started from. These, upon approaching, were found to be about a dozen young hair seal pups, of from one to ten, to fifteen days old, not sufficiently strong to have moved out of his way, laying closely huddled up by the side of the rocks.

Our party having learned what were not fur seals, and wisdom enough to engage no more sea-lions, after skinning our dead lion, and taking with us his skin as a remembrance of our hunting tour, took up the line of march, and returned to the brig.

By great exertions, a full cargo of fur seal skins was procured for the brig, by the month of January; at which time, too, it became very evident, Captain Steele was in no haste to return.

This sacrifice of time was more than I felt willing to submit to, and there offering, very fortunately at this time, a passage to New York, with an old acquaintance, I obtained Captain Steele's consent, and embraced the same, where we arrived in March.

The *Betsey*, after visiting the West India Islands, arrived in the month of June following, and as soon as she was made fast to the wharf, Captain Steele was informed that the owners had no further service for him; Mr. Whetton taking the charge.

Being absent from the city, upon my return, Mr. Elias Nexsen, in behalf of the other owners, having informed themselves of the occurrences at the Falkland Islands, had directed my full proportion of shares to be handed me; the same as would have been the case had I remained with the brig throughout the voyage.

CHAPTER III

VOYAGE TO ENGLAND

IT is no doubt fresh in the minds of many readers, that the distress of the people in France, during the French Revolution of 1793-9 (for bread and bread stuffs), was very great, and as a necessary consequence, of such scarcity, that prices were accordingly high. This well-ascertained fact was a sufficient inducement to a people situated as were the Americans, and possessing all their enterprise and spirit of adventure, to undertake supplying in part, the demand for these articles, although the hazard attending the same was rendered immensely great, from the many English cruisers that then swarmed along every portion of the coast of France, and whose watchfulness made it almost impossible for any vessel to escape them.

It was at this period that the new ship *Portland*, (Captain Thomas Robinson, a very able seaman, and afterwards in command of the United States' frigates *New York* and *John Adams*), belonging to the house of Bowne & Eddy, merchants of New York, lay at that port, on board of which vessel the author entered as first officer. The *Portland* was bound to Norfolk, there to take in a cargo of flour for France. She sailed from New York early in May, and four days after arrived at Norfolk, where she received her cargo; things were in this state of forwardness when a drawback was put upon them by the desertion of all the hands, who had immediately after, gone into the country; but occasionally came in town, as appeared from information afterwards obtained, and were in the habit of passing their time at a boarding house on the side hill; to this house, besides the front door at the lower story on the street, there was another, leading by a long wooden stoop to an alley, which last was used as a security against surprise, for, when alarmed by any unwelcome guest at the front door, it was by this second story that they would retreat.

One evening, hearing that the seamen were in town, after obtaining the assistance of two police officers, we repaired to the house, in order, if not impossible, to bring them back to duty. The leader of this desertion was one Wright, an old seaman, with whom the author was a great favorite; knowing this, Captain Robinson directed him to go round to the door in the rear, and there stand sentry, observing, "They are, no doubt, led on by Wright, and your influence with him will have much weight in bringing about the desired result, should they strive to escape that way; the seamen, being alarmed by the captain and police officers, at the front door, as was conjectured, attempted a hurried retreat at the upper door, under Wright's direction, but seeing me with the boat's tiller over my shoulder, Wright turned suddenly round to his shipmates and called out, "Here is Mr. Fanning, with the holy stone (a large stone fitted with a handle, similar to a scrubbing brush, for scouring decks, and only known by this name among seamen), on his shoulder; we may as well surrender." I embraced this fortunate opportunity, and remarked, "You know, Wright, that I am your friend, why do you conduct thus?" And, in answer to his inquiry whether I would make peace with the captain, I assured him and them that the captain would not "tighten their rigging" for them, they, therefore, consented to go on board ship again, where, for the remainder of the voyage, they cheerfully did duty to the full satisfaction of the captain. The ship now dropped down to Hampton Roads, and on the next day put to sea, bound for Havre, in France.

We had fine weather during the passage, without anything worthy of note occurring, until, on soundings abreast of Scilly Isles, on which day, early in the morning, a moderate and fair breeze blowing at the time, accompanied with a smooth sea, our ship was fired upon by a British private armed cutter, which, upon our bracing the maintopsail to the mast, and laying the ship by, came under our stern and hailed; "Where are you from? Where bound to?" to which interrogations he received the reply of, "From Norfolk, bound to Havre de Grace." "Then come immediately on board of me with your papers," demanded he of the cutter; "if you wish to examine my papers, you must come on board my ship," calmly replied our captain, "for I cannot consent to leave her." "You d—d Yankee, hoist out your boat this instant and come on board, or I will sink you!" continued the commander of the cutter. "I understand my duty too well," answered Captain Robinson; "and your vulgar threat will not sway me from it!" The last bravado from the cutter was succeeded by a discharge of musketry, the balls whizzing over our heads.

Ever careful of the lives of his men, Captain Robinson directed the topgallant sails to be run on the cap, and every man sent below, adding, "Mr. Fanning you had better follow, for I have a desire to know if this boasting commander will dare to carry his threat into execution;" I, however, declined; observing, that I preferred remaining at my station and take my chance with him. After very handsomely peppering our sails and upper rigging, which showed their pieces to be well elevated, the tompions of their guns were taken out and they made to bear upon the ship, which manœuvre was followed by a threat, couched in the same courteous language as its predecessors, but producing no other notice than the deliberate reply of Captain Robinson, "That if he desired to examine the ship, it was necessary to come on board for that purpose, for he would not so far leave his duty as to obey the order; more especially," added Captain Robinson, "as it contains so much of the fine feelings of a gentleman and officer under the flag of His Brittanic Majesty." The firing here ceased, and an officer from the cutter boarded the ship, and reported her as being loaded with flour, and bound for Havre; she of course was a prize, and having a prize master with fifteen men placed on board, who took possession of her, was steered for a port in England.

When off Falmouth, we were fallen in with by a British frigate, who, as the most summary mode of procedure, sent a shot athwart our forefoot, and brought us to. A boat from the frigate, with seamen and marines, immediately boarded the *Portland*, and, with the exception of the prize master, sent the privateer's people to the frigate; laying the course of the *Portland* for Falmouth, at which port she anchored the next day. Here the cargo was subjected to an examination, and after being appraised by two merchants, the one chosen by the collector of the port, and the other by Captain Robinson, and according to their appraisal paid for, was stored in the king's storehouse: yet in order to obtain the payment, it became necessary for Captain Robinson to make a journey to London, with a certificate of proceedings.

The day after the *Portland's* arrival in Falmouth harbor, it was requisite for the captain to take his first officer and two of the crew, together with the ship's log book, before his Honor the Lord Mayor, for the purpose of undergoing an examination in relation to the ship and cargo. Upon the mention, by the clerk of the court, of the first officer's name, his Honor, looking earnestly at him, says, "Fanning! pray, Mr. Fanning, let me ask, are you a relative of General Edmund Fanning, the Lieutenant Governor of Prince Edward's Island?" I replied, "I am, Sir, and have the honor of receiving my name from that uncle." "Indeed, my young friend," his Worship was pleased to observe, after a cordial shake of the hand, "Your uncle was my old and long tried friend, to whom I am under many and weighty obligations; and you cannot better please me, than by making my house your home during your stay at Falmouth." At the same time giving Captain Robinson an invitation to spend the evening with him, which was accepted; after completing the investigation, the party were suffered to return on board ship.

The next morning, Captain Robinson, while relating the occurrences of the preceding evening, mentioned the particular inquiries of the Lord Mayor in reference to his (the captain's) first officer; saying, he knew I had got a weighty friend in his Worship, and that it was a fortunate hit for himself, as he believed he would be of no little service to him in his present difficult business about the ship and cargo; adding "He has already given me liberty to mention any favor wanted, and has insisted upon, and obtained my promise, to dine with him and a few friends on the following Wednesday, in company with yourself, that he may have the pleasure of introducing you to his friends; therefore," continued he, "prepare for that day." This last caution was highly needed; for to be thus without the least previous notice, taken into such favor, was an honor as totally unexpected, as it was for me to know what next was to come. However, encouraged by the captain, and thinking it best not to shear off from, or push aside any good fortune, I prepared for the day set, by "storing and stantioning" my mind and fortitude sufficiently well to get through the dinner campaign with credit, and endeavor, if possible, not to commit any blunder. The important day and hour having arrived, in company with Captain Robinson, I repaired to the hotel, where we were received by his Honor the Mayor, with a friendly shake of the hand, and smiling countenance, by whom we were kindly introduced to his friends, in number about thirty, each wearing a large powdered white wig, and bearing upon their persons marks of royal favor, badges of honour, &c. Dinner being announced, the company forthwith proceeded to the apartment where the same had been prepared. It was a rich entertainment, and every thing to be desired was there in great plenty; in truth, the table was, through the several courses, loaded with the best the Falmouth market afforded, after which succeeded the desserts. Seated, by his Worship, at his right hand, with Captain Robinson at his left, so high, and in such company, the author, then but a green Yankee, and totally unexperienced in the forms of such society, felt like being in the wrong place, or, in other words, as "a fish out of water," and with something akin to pleasure welcomed the breaking up of the party, although to please his new friend was his most earnest wish, and any thing in his power would most cheerfully have been bestowed, to add to the general good feeling that prevailed during the dinner.

It was, in compliance with the Mayor's pressing invitation, presented on the breaking up of the dinner party, that the next day found me, punctual to the hour, at his office; he was alone, and after some conversation in relation to the general, my uncle, observed, "From what Captain Robinson has given me to understand, in regard to your history, my young friend, I am very anxious to be of some service to you; to do something that I am sure will be gratifying to my old friend, the general; for I have often heard him speak of you. Now suppose," continued this kind man, "that I should use my exertions to procure you a commission, as lieutenant in his Majesty's navy; you shall be under no obligation, for I can readily, and will most willingly, do it;" accompanying this offer with that of funds, to enable me to move with credit to the rank, until the voluntary task should be taken from his hands by his friend, the general; "and whom," said he, "I know will be greatly pleased with the proposal, and more so, to learn that it was carried into effect."

In reply, I observed to his Worship, that I was at a loss for words to express my thanks for his goodness, but having a wife and child in America, being a republican in principle, and too much attached to liberty and my country, to enter the service of his Majesty, I must decline; and these reasons must plead in excuse for my not accepting his very parental and unmerited proposal; yet it should always by me be held in grateful remembrance.

At this, his Honor, taking me by the hand, replied, "My young friend, notwithstanding you decline my proposition, I like you none the less for your candor, and should you ever stand in need, rely upon my friendly aid; and I again insist upon your always considering my house as your uncle's, and your own home." To these several specimens of good old English hospitality, I could but give my warmest thanks in return, and shortly after took leave.

There is a pleasure in recounting and reviewing these tokens of sincere friendship: they bear their testimony to the truth of the existence of the same, and are a sort of rallying point, around which the best feelings of our nature collect, preparatory to again encountering the selfishness which, unfortunately, characterises the dealings of man with his fellow-man. In this particular case, it was pure disinterestedness on his part, for there was no personal advantage, or private gain, to be obtained by this attention; the good understanding between this worthy man and my uncle, being the sole cause; and this would have remained the same, had no attention been shown to me.

Captain Robinson had not yet returned from London, whither it will be recollected he had gone, to obtain pay for his flour as appraised, when the British frigate *Nymphe,* Sir Edward Pellew, arrived at Falmouth, bringing in with her as a prize, the French frigate *Cleopatra,* which vessel she had captured, after a severely contested action.

This event, as might well be expected, caused great exultation, and the most lively joy was manifested at this additional trophy to the prowess of their countrymen, by the townsmen; the frigate, too, was again repaired, and made ready for sea, with the utmost despatch. It was necessary, however, to have more hands, and for this purpose a hot press took place: seamen were taken from all the vessels in port; from the American vessels as well as others. The second mate, carpenter, and all the hands, excepting the steward and cabin boy, were taken from the *Portland.*

In this dilemma, I repaired to Mr. Fox, the American consul, to inquire how to proceed, in order to regain at least a portion of our men, all being, to my knowledge, American citizens, but two Swedes. Here were collected the captains of several American vessels on the same errand with myself. The consul said, the men had been all taken to man Sir Edward Pellew's ship, who was anxious to have the credit of putting to sea, if possible, in the shortest time on record,

THE CAPTURE OF THE FRENCH FRIGATE "CLÉOPÂTRE" BY THE BRITISH FRIGATE "NYMPHE," CAPTAIN EDWARD PELLEW, ON JUNE 18, 1793

From a colored aquatint by Cornelius Apostool after a drawing by Lieut. T. Yates, London, 1794

THE BRITISH FRIGATE "BLANCHE" BLOCKADING NEW YORK HARBOR

From an engraving in the *Naval Chronicle* (1814), after a drawing by Pocock

after such an important and brilliant victory, in order to surprise his enemy; and standing now, as he did, so high in favor with the government and the people, it was requisite to be patient, moderate, and judicious, in our movements: concluding with a request that we would call at his office the following day, when he would endeavor to see Sir Edward, who was in the habit of landing daily from his ship, at eleven o'clock A. M., at the pier directly in front of the consul's office, and must pass up the street by the door, and he would then speak to him upon the subject.

The next day, while conversing at the office upon our grievance, and keeping a lookout, Sir Edward landed as usual, accompanied by three other gentlemen, to whom Mr. Fox advanced, and politely bowing, lifted his hat, at the same time mentioning the subject of the impressment of our men. He was answered with coldness, receiving in return to his salutation, simply a move of the head, and the remark, that want of time rendered it impossible for him (Sir Edward) then to stop; leaving Mr. Fox at liberty to return to us (in waiting). You see, gentlemen, said he, I can do nothing for you, until time shall a little cool the high feeling resulting from this victory; adding, that if we thought fit to call at his office daily, so soon as a favorable opportunity occurred, our business should be attended to. We did call daily, receiving much the same answer, with less prospect of success on the second and third days, if there was any alteration, than appeared on the first. These continual fruitless efforts caused some of the captains to scold at Mr. Fox very bitterly. While so unfortunate a state of things weighed very heavily upon my young mind; having been left with orders to have the ship ready for sea by the time Captain Robinson returned from London.

It now occurred to me, that in all probability my good and worthy friend, his Honor the Mayor, was acquainted with Sir Edward, and would be disposed to favor me with his assistance. Therefore, without mentioning my intentions to any person, I repaired to the Mayor's office, and found him busily engaged writing; he arose, and after a shake of the hand, requested me to be seated. I mentioned the difficulty that had been brought upon me, and that my present visit was for aid and advice. After giving a statement of the transaction, in reply to his question, what difficulty it was, "is that all?" said he, with a smile, "Oh, then I think in case you can keep your own secret, a plan can soon be put into operation, that will very shortly restore your men." After being informed that all of the *Portland's* crew, but the two Swedes, were Americans, he remarked, that "Sweden being in amity with his Majesty, should they volunteer and enter, no authority could bring them back to your ship; but as to the others, I presume there will not be much difficulty in regaining them; for," continued he, "I am intimately acquainted with Sir Edward Pellew; he is one of my most particular friends; to him, Mr. Fanning, I will give you a note of introduction; his residence is a conspicuous establishment, a little up the river, on the opposite bank; it is very well known, and any person will point it out to you from among the other seats; tomorrow morning, between eight and nine o'clock, take this note to him; before he leaves town, I will endeavor to see him, and explain your case; let me see now," said he, "how well you can manage an embassy with these credentials."

The next morning, manned the ship's yawl with all hands, viz., steward and cabin boy, and at the appointed hour, proceeded to the landing abreast of Sir Edward Pellew's seat, the same having been pointed out to me by the custom house officer on board the ship. Upon ringing at the gate, a porter appeared, who stated Sir Edward to be at home, but then engaged at breakfast, asking my name, attended with a request that I would call again; I desired him first to deliver the Mayor's note to Sir Edward, and then he would get an answer. In a few moments Sir Edward came to the door, having the note in his hand, and very politely inviting me, if I had not breakfasted, to step in and take a cup of coffee. Upon my declining, he observed, "Your business, Mr. Fanning, I perceive to be the *Portland's* men; are they all Americans?" I replied, they are, all except two Swedes. "Do you, Mr. Fanning, know that to be the fact,—at least, will you give me your word of honor you believe such to be the fact." I answered, "I do"; and then mentioned, that to my certain knowledge, the second mate had a wife and three infant children dependent on his earnings for their subsistence. At this moment, the lady of Sir Edward, who, upon perceiving him receive a note from the porter, and instantly retire, conceived something more than ordinary was about to take place, and had therefore followed him, unobserved by himself, until she placed her hand with much affectionate earnestness upon his arm, exclaimed, with much sympathy, "Oh, my love, do not take that man from his family!" "Hush," he mildly replied, "do you, my dear, go in and finish your breakfast, and leave to me my ships affairs, for be assured I understand the management of them thoroughly." Then turning, he continued, "Mr. Fanning, at one o'clock call on board the frigate; if I am not there at that hour, the commanding officer will have my instructions in reference to the *Portland's* men." I thanked him, and returning to the boat, went on board ship, there to wait with the greatest patience for the appointed hour.

At one o'clock, I went with the yawl alongside the frigate: while going up the gangway, was met by a midshipman, who inquiring the name, and being answered, "Fanning, from the American ship *Portland*," requested me to wait until he should pass the same aft. Shortly after, a lieutenant made his appearance, who very politely led me to the quarter-deck, where were gathered a circle of the officers of the ship, in the midst of which was the lieutenant then in command, to whom I had the pleasure of an introduction. On inquiring whether Sir Edward Pellew had left any directions in reference to the *Portland's* crew, he stated that with the exception of the two Swedes, who had voluntarily joined the frigate, all should be returned. "The Swedes," he added, "are subjects of a nation in amity with our government, and it would be more than the Lords of the Admiralty would undertake to do, to allow them to be taken from on board."

Exceedingly well pleased to come off thus with the

majority of our men, I mentioned that nothing unreasonable was desired, or meant to be asked for; and while partaking of the refreshment afforded by some crackers and cheese, the whole, "sailor like," accompanied by good punch, an order had been given to embark the *Portland's* crew, with their luggage, into the boat. Matters generally are conducted in such an orderly and clock-work style, that very little ado is made on board a man-of-war while such business is being attended to: they accordingly were soon reported as being all ready, when after receiving a friendly invitation to call on board the frigate again, I left them, and pushed off from alongside.

The joy of the mate and men at being thus once more free, was warmly expressed in that plain matter-of-fact style so peculiar to seamen.

The mate said it was perfectly plain that the tide was setting in their favor, from what had occurred the evening previous on board the frigate; for soon after Sir Edward arrived on board, an officer came to Wright, attended by a midshipman, and directed him (Wright) in future to mess with the middy, while at the same time a quarter-master was ordered to have the *Portland's* men placed in his own mess, and to see that full allowance was dealt out to them; "and this morning," said he, "an officer mustered and examined us as to what countrymen we were," &c. It was at this time that the two Swedes requested to enter, and were accordingly placed on the ship's list.

In calling at the consul's office the following morning, I found the several captains as yet quite unsuccessful in obtaining even a shadow of hope that their cases would receive any attention from Sir Edward. Nothing would answer, after they understood that the *Portland* had all but two of her men back, but they must know how, and in what way, it had been managed. Considering myself under very strong obligations studiously to avoid discovering the manner of procedure, and to give the best face to the whole, I answered, that after going on board the frigate, I plead how hard it was for me a poor mate — the captain being absent — the ship to be got ready for sea against his return — that nothing could be done with only a steward and the boy — and the like urgent reasons, so successfully, that the officers gathered around me, talked, laughed, and guessed I was a true Yankee; and after a little such chat, one of them gave orders to have all my men returned, except the Swedes. The consul was much surprised while this was being related, and said, he could not account for its successful termination. After listening very patiently, some of the captains began to bear down hard upon him, saying he was wanting in his duty to his country, to his impressed fellow-citizens; charging him almost with being the cause of their present distress; that all this should be made known at Washington, &c. &c. The consul was silent a few moments, but finally, taking his hat, said, "Gentlemen, if you think I am neglectful, or deficient, I have only to say, the same road that Mr. Fanning has taken, is open to you all; my duty I shall attend to;" and immediately left the office. Not wishing to be farther questioned, I soon followed his example.

While we lay here, the miners employed in the vicinity, and also at Cornwall, in consequence of their very great distress for the want of bread, and other necessaries of life, had risen in large numbers, and were marching upon Falmouth, with the intention of seizing upon the flour, amounting in all to many thousand barrels, then in store at the King's store-house, which, like our own, had been captured while destined for France, and there placed for keeping. The citizens were called upon to arm themselves instantly, and assist in the defence of the place. Troops under arms, were held in readiness to defend those points most liable to an attack, and a slight fortification had also just been completed at the bridge, which crosses the river here, as the advance body of these miners was seen moving forward along the road leading to it, in numbers about fifteen hundred, and from appearances, determined upon storming and carrying it. A white flag was however displayed, and a parley had between their leaders, and the city authorities, at which a price was agreed upon for the flour per barrel; they (the miners), to take any proportion they chose, on paying for it at the store-house. By this arrangement, peace was once more restored. The miners, after receiving some hundred barrels, took up their return march for the mines. Many of them, it was credibly reported, and generally believed, had not been, previous to this affair, above ground for several years; to judge indeed from the appearance of some of them, one would not have been supposed to be greatly in error, to assert they were the choicest of Pluto's myrmidons.

Captain Robinson having returned from London, and there being no farther cause for delay, we soon had the ship ready for our homeward voyage; there were a few cabin, together with a large number of steerage passengers, engaged to go out with us. These last were mostly in families; and so soon as their various domestic and farming utensils, their provisions, and all their most needed articles for the new country, were duly stowed, we sailed for New York in ballast; the passage was unattended by anything more than that of usual occurrence, until our arrival at the eastern edge of the Grand Bank of Newfoundland, when, at about three bells, A. M., (half-past one) it blowing a violent gale, and a heavy sea running at the time, our ship then laying-to under a reefed fore-course and mizen stay-sail, was struck by a mountainous sea, which broke on board, and knocked the ship over on her beam ends in such a manner as to bring the lower yard-arms at the leeward in the water, shifting a part of the shingle ballast over to leeward, the water at every lee lurch making its way down the gangways; our stay-sail also was split and torn to pieces by the force of this breaker. As soon as the stroke had passed, all on deck at the time were striving to secure the safest place, at the weather side of the ship. I had gained a station on her side, by the main channel, Wright, the old seaman, by my side; when, observing by the force of the gale on the upper or weather yard-arm of the fore-course, that the ship's head was paying off from the wind and sea, the thought occurred to me that if we could now but humor her with the fore-sail, she might go around on the other tack, and yet save us. Bidding Wright to follow, we crossed by the fair leader from the main rigging down to the main-mast, by which the fore-braces lead; here I directed him to cast off the weather-brace, and pass it to his shipmates on the other side, to haul in upon; at the same time, by gradually loosing the lee one, (the lee yard-arm and lurch of the fore-course dragging in the water) kept the ship going off around before the wind, and as she went off partially righting, the helm then being put hard a-weather, and getting headway on her, she fortunately went round, and came to on the other, the larboard tack, with about two streaks of her deck under water, on the now weather side, with such a rank heel to windward, against the gale, as enabled those below to come on deck.

At this moment, Captain Robinson, started from his rest by the noise and confusion, appeared seeking anxiously to know whether all was safe. He was answered, the only loss was the mizen stay-sail; when, after looking around, he ordered the fore-sail to be trimmed, to bouse taut the fore-braces, brace all the yards about, and all hands, except two seamen, were sent below, to ascertain the quantity of water in the ship's hold, and to endeavor as quickly as possible to turn over and secure the ballast. This last operation we completed, by lashing amidships, bights of cable, and spars, fore and aft, to the stanchions, so as to prevent the ballast from shifting again, and then trimming it over to the starboard, until the ship was brought "upon her legs," when, to our great joy, the pumps soon freed her.

The awfulness of the night, in addition to the heaving of the ship, and the repeated gusts of wind which continually swept over our deck, together with the shrieking and crying of the women and children, made these few hours (so nearly fatal) to be the most wretched of my life. Nothing could exceed the terror of the steerage passengers during this gale: many of the principal, and most aged, were devoutly engaged at prayers, not expecting the ship could hold out through the storm. Some difficulty was experienced in persuading them (while we were proceeding to regulate the ballast) that the danger was nearly over: their young men, however, readily consented to assist us while busy below, and worked most faithfully, as men usually do, when their lives, and all they possess, are at stake.

Shortly after, the storm began to abate, and twenty-four hours time, we were favored by a moderate and leading breeze. Three days after, we had soundings on the Grand Bank, where we caught a few fine cod: these excellent fish came very acceptable to our passengers, as well as to the ship's company. A few days more of making and taking in sail, concluded the voyage, and saw us safely returned to New York.

CHAPTER IV

VOYAGES TO THE WEST INDIES

IMMEDIATELY after the return of the *Portland*, as related in the chapter preceding, through the kind attention of those friends for whom I had previously made several voyages to the West Indies, I was, for the first time, placed in command as master, viz., on board the new schooner *Dolly*, belonging to the house of Snell, Stagg & Co., of New York. In the Curacao trade, for which this vessel was intended, I was perfectly at home; experience acquired during former voyages, had given me many useful lessons, that would now be brought to bear.

While exercising a master's usual privilege, of having several tons freight free allowed him, I had stowed the same with some eighteen dozen beaver hats, in cases, together with one hundred and sixty kegs of butter. It was a most fortunate venture, for a beginning. The latter article especially, the Curacao market was greatly in want of; in fact, there was none to be had until the *Dolly's* arrival, at which time it readily brought fifty cents per pound, weighing the firkin as discharged from the vessel, no deduction or allowance for tare being required. The huckster women, even at this price, were continually pushing and wrangling with each other, for the firkins, as they were passed up from the hold, and keeping up an incessant clatter all the time, each most stoutly insisting upon being first served. This butter cost but twelve and a half cents at home; and the hats, at invoice price, two dollars and a half each, produced five, without even so much as being unpacked. They were sold to some Spanish traders, merchants from the Spanish Main.

The schooner, with a cargo consisting principally of bread stuffs, was in a few days, in company with a fleet of some thirty or more sail of all riggs, got under way for Sandy Hook, there to discharge our pilot. Among this little fleet was the ship *Portland,* Captain Robinson, my late commander, with whom, by means of the speaking trumpet, a seaman's hearty wish for each other's future health and prosperity was exchanged. At sunsetting of the same day, took our departure from the highlands of Neversink, and after passing through the continued changes of winds and weather, which occur on a West India passage, we arrived at the port in the island of Curacao, on the twenty-fourth day out.

This is a most excellent harbor, and I believe it to be one of the snuggest and safest among the whole of the West India Islands. It opens to the south; the entrance is narrow, and somewhat difficult to square-rigged vessels. The attendance, however, at the fort, on the weather chop at the harbor's mouth, is so good, and prompt in rendering assistance, by means of men, boats, and hawsers, that it seldom happens vessels get on shore. Such as are bound here, may therefore stand boldly in, under full sail, gaining all the headway possible, until they come to luff in, so that if possible, they may shoot quite into the harbor, before the sails take aback, at the same time letting all run, and clewing up; if the harbor has been gained, the sails will then become becalmed under the lee of the castle and buildings of the town: yet vessels making this island towards nightfall, whose masters may be unacquainted with the manner of entering, had better lay to, because of the low island stretching towards the east end of the main island, at about one league distance; and the strong current always setting to the westward, along this coast, especially at this island, makes it very difficult to fetch up to the port, should they unfortunately fall past it: at the same time by laying to, the early part of the following day can be had to run in, directly from the mouth of the bay.

The market here is in general very good, and well supplied with meats, such as beef, pork, mutton, &c.; also a variety of most excellent fish, with the best of that delicious dish, the green turtle. This celebrated dish I also found at Bermuda, though not equally well cooked, and not of so rich flavor there, as they give it here. The manner of cooking the callipatch and callipee here, is also superior to any I have ever met with; yet the market of Bermuda is certainly well supplied with fish, generally, and an epicure would there find his nicest wish accommodated. Their "grouper," especially, is a most excellent dish, and the inhabitants insist upon its being the most so of any in the world. Here, too, are to be had plenty of vegetables, fruits, melons, &c. &c.

This place is greatly supported by its commerce with the Spanish Main; and if good policy is used, this will and must continue to be the case, both from the excellency of its harbor, so perfectly safe and commodious, and the great and inviting advantages it presents for the repairing of vessels, &c.

On one of the *Dolly's* voyages to, and while at this port, it was blockaded by the British, who at that time were capturing all American vessels, and sending them to New Providence, Bermuda, Halifax, &c., for adjudication and condemnation. The *Dolly* left the port in the night, and keeping the island shore close on board, until its west end was doubled, fortunately evaded the blockading squadron, and got safely through the Moona passage. Everything succeeded well, until abreast of Cape May, where we were boarded by a Philadelphia pilot, who informed us, that the British frigate *Blanche,* Captain Faulkner, was off Sandy Hook making prizes, and sending to Halifax all vessels from the blockaded ports; yet gave it as his opinion, that the late heavy gale from the north-westward, must have blown her off to the eastward, so that by working along the Jersey shore, with the then fine breeze, our chance was good to arrive within Sandy Hook before the frigate could regain her station. We accordingly worked to the northward, until abreast of Barnegat, keeping the shore all the while close aboard; but while endeavoring to weather the shoal (Barnegat), a sail was descried in the northeastward, and soon discovered to be a man-of-war. Every exertion was now made, by tacking in shore, &c., to get clear of her; but her superiority in sailing was so great, that she soon passed our weather beam, tacked, and bore down upon us, and when nearly within hail, fired a swivel shot just ahead of us. We now hove to, and were soon accosted with the usual interrogations, on such occasions, as "Whence come you?" "From what port?" and the like; he receiving in return the most satisfactory replies. The schooner was shortly after boarded by an officer from the frigate, accompanied by seamen and marines, the officer remarking, as he came on board, "A good prize, sir! We shall alter your course for you, and take you to Halifax;" adding, "Captain Faulkner desires you to take all your papers and the log book, and repair with them on board the frigate." In conformity with this requirement, I stepped into their boat, and soon was pulled alongside the frigate, where, upon gaining the deck, was met by an officer, habited in a short roundabout jacket and white trowsers, whom at the time I supposed was a midshipman, sent to show the way to the captain. "Good morning," said he, (by his manner apparently anxious not to add the least unkindness to my loss of property) "Come, walk below;" and led the way down the gangway, then aft into the cabin, the door of which swung to after our passing in, by a weight. It was now very certain that this was none other than Captain Faulkner himself. Having taken a seat at the table, he requested me to follow his example, and immediately proceeded to examine the schooner's log book. This, like the log books of most, if not all merchantmen, had the names of the vessel and master, written in a large hand, and very distinctly, on the covered leaf; for turning it back, he saw it, and inquired whether I was any relation of General Edmund Fanning, now residing in Portman Square, London. I replied, that I was a nephew of his. He instantly closed the book, and without touching or looking at one of the schooner's papers, said he had no doubt every thing was correct; expressing a wish, as a steward appeared in answer to his call, bringing in some wine, to drink to the health of the General, whom he gave me to understand was his particular friend at Court, was his godfather, and had given him much assistance in his education and advancement in life; "I am, therefore," continued he, "very happy to have fallen in with you, though grieved to be the cause, while performing my duty of one moment's delay to your vessel;" and concluded, with observing, that as he presumed I was anxious to see my family and friends, we would therefore walk on deck.

After some conversation between Captain Faulkner and the lieutenant, (the same that had brought me off to the ship) the latter proceeded to the boat alongside, where, also, after receiving a hearty shake of the hand from the captain, with a hope that I should find all friends well upon my arrival at New York, I followed. As the frigate's boat reached the schooner, the lieutenant made known to the officer then in charge, that himself and men must now return to the ship. The latter, however, did not appear disposed to quit so good a prize in very great haste, and began, perhaps in expectation of removing the erroneous impressions that he had no doubt Captain Faulkner lay under, in thus easily returning the schooner, by answering that she was from a blockaded port, and had, moreover, a very valuable cargo of coffee and hides on board. The matter, nevertheless, was speedily settled by the former lieutenant's decisive reply, that it was the captain's orders. With no small degree of disappointment, strongly depicted in their countenances, both officers and men now prepared to leave the *Dolly*. Not willing that all the courtesy should be upon one side, and as some small return for Captain Faulkner's great attention while on board his ship, I had a bag filled with oranges and limes, in fine preservation, put into the lieutenant's charge for him, each of the officers receiving a small lot for himself and mess, previous to leaving the schooner. Again at liberty, we made sail for New York, and while passing the ship, Captain Faulkner stepped aft and acknowledged the above compliment.*

The schooner's safe arrival, caused considerable surprise on 'change, and among our acquaintances, as very few vessels had yet been able to escape the frigate, and many speculations and shrewd guesses were also advanced, to account for the same. It was a little singular, that the evening after the examination and giving up of the *Dolly,* the frigate captured the *Portland* (Captain Robinson still on board). She was from one of the blockaded ports, with a cargo of coffee, &c., and sent (by Captain Faulkner of the frigate) to Halifax, where both vessel and cargo were condemned.

*He was (as I have subsequently been informed) killed in a desperate engagement with a French frigate, in the West Indies, while in the act of lashing his opponent frigate's bowsprit to the capstan of his own ship, and just at the moment of victory.

CHAPTER V

FIRST VOYAGE ROUND THE WORLD

IN the early part of the month of May, 1797, it was the good fortune of the author to meet at New York with Captain John Whetten, a gentlemen distinguished as an able navigator, and at the time in command of the ship *Ontario,* in the China trade. With him originated, and with him also was the project first discussed, of fitting out a suitable vessel which should proceed to the South Seas, there to procure a cargo of fur seal skins, and with this cargo thence to cross the Pacific for the Canton market, where the article was well ascertained to be greatly in demand, and held at prices that furnished good grounds upon which to hope a very handsome profit would be realized to those who were disposed to engage in the business. Another great inducement held out in favor of the attempt, was the probability that Captain Whetten himself, in the *Ontario,* would be at Canton at about the period of the arrival there (which would be in our fall part of the year, say the month of September, October, or November) of any vessel shortly fitted out. The intimate knowledge that Captain Whetten possessed of the manner of doing business with those people, and the great assistance he could afford, being conversant in all their intricate trade, and in the purchasing of silks and other articles for the New York market, as our homeward cargo, were certainly such arguments as were well calculated to increase the confidence of success in the contemplated voyage. This was an opportunity not to be left unimproved; and to one naturally possessed of an ambitious and aspiring mind, with a strong attachment to a seaman's profession, increased as it had been, since my first visit to the South Seas, by a perusal of the voyages of such circumnavigators as Drake, Byron, Anson, Bouganville, Cook, and others, the hope of being able to add some new discoveries to the knowledge already in the possession of man relating to those seas, and the no less flattering hope of realizing a fortune, if the enterprise were well conducted, and successful in its termination, were sufficient to bind me to exert myself in bringing about this desired voyage.

Every view was encouraging; but funds were necessary, and to raise these without delay, I applied to that upright and liberal merchant, Mr. Elias Nexsen, with whom also to consult and advise upon the best means of securing the early fitting out and sailing of the enterprise. To the information and encouragement given by Captain Whetten, was added my own strong confidence in its practicability, and the flattering results that such an adventure held forth. The plan met with his entire approval, and after some conversation with Captain Whetten, on 'change that day, in the afternoon of the same he made the offer of his brig, the *Betsey,* then in port. She was New York built, a little short of one hundred tons, and an excellent vessel of her class. "If she will answer," said he, "I will put her into the business, and at whatever price, upon a minute inspection, her value shall be ascertained to be, I will take the one half in the adventure of the vessel, and her outfits.

The *Betsey* had been built under the superintendence of Captain Motley, for a Charleston packet, who was also at the time a prominent owner in her. Her good qualities as a first rate sea boat, &c., from having been several voyages in her, I well knew: her size was the only objection; a little larger, and she would have been the very vessel; however, this was but an experiment, it was on a new trade, and should it eventually prove unfortunate, then she was abundantly large. I was unable to take more than one eighth myself, but the remaining three were by the evening of the same day taken by other friends, and thus the whole amount required to insure the sailing of the vessel was made up. An inventory was taken, agreeable to the understanding at the commencement, by which the value of the vessel was ascertained, and made satisfactory to all concerned.

The brig was next hove out, and thoroughly examined, previous to receiving anything on board. Her rigging was completely refitted, and put in the best possible order. The stores, provisions of every kind, with a small invoice, consisting of beads of different sorts, small looking-glasses, buttons, needles, cutlery ware, and the like, suitable for trading with the native Indians, of places that we might visit, were put up in the best manner for preservation, and part of the crew shipped.

Among these was one, to all appearance, as innocent as he was simple. A tall, raw-boned, stout, good natured looking fellow, who, as he said, had just come down the North River; one from among the Green Mountain boys. In name, so was he in character and person, a true, a real Jonathan, to all intents. Folks had told him, he said, that my vessel was going t'other side of the world; he had never seen the salt water before, and he would now like to have a chance in her, as he wanted, most nationally, to see how t'other side looked. He guessed he knew he could please, if hard work could do it, for he had come to seek his fortune; and throughout the voyage, he was found to be willing and obedient. In fact, his simplicity and readiness to do whatever would be required of him, made it necessary that care should be taken that no imposition should be put upon him, either by the officers or men.

In less than a month from the time that the information, on which the present voyage was fitted out, had expired, did the *Betsey* sail from New York, to stop at New Haven (the native place of Mr. Caleb Brintnall, the first officer, a great disciplinarian), and afterwards at Stonington, to obtain and complete her compliment of men, in all twenty-seven; for it had been concluded to be the most judicious policy, to select the greatest proportion of them from the New England states.

CHAPTER VI

PASSAGE TO THE CAPE DE VERDES

HAVING obtained the number of hands required for the *Betsey*, on the thirteenth day of June, 1797, she was got under way, and proceeded to sea, from Stonington, Conn. When off Watchhill point, (situated about nine leagues to the northward of Montaque light, on the east end of Long Island) she was brought to, in order to discharge the pilot, and the occasion was embraced, as the best suited to ascertain the minds and inclinations of the seamen. All hands were therefore mustered on deck, aft, and liberty was given to all such as were disinclined to proceed on the voyage, to all those who were unwilling to encounter the dangers, privations, and sufferings, usually attendant on similar expeditions, now to return with the pilot. Notwithstanding this, no one seemed so inclined, but all, to a man, answered, their desire was to proceed on the voyage, confirming the same by three hearty cheers. And here it may be remarked, that a more orderly and cheerful crew never sailed round the world in any vessel. The pilot accordingly returned by himself.

At six P. M., we took our departure from Block Island, with a fine breeze from the southwest. The next day, at meridian, our latitude was 40° 48′ north, with land, supposed to be Martha's Vineyard, then in sight off our lee beam. The wind now veered to the S. S.

VIEW OF NEW YORK HARBOR IN 1796
From an engraving by St. Memin

VIEW ON THE PATAGONIAN COAST

From an engraving in Anson's *Voyage Round the World*, London, 1748

W., accompanied with hazy, but not unpleasant, weather. Some lines were fitted up by the officers, and with these they succeeded in catching several hundred fine mackerel: saw at the same time a number of sperm whales, and vast shoals of what are known among whalers as blackfish, and also many herringhogs.

On the 16th, at noon, our latitude was 39° 28' north, longitude 65° 42' west. Weather very fair, with a fresh gale from the S. W. This day saw numbers of the fish called albercores, around the brig.

25th. Latitude at noon, 39° 22' north, longitude 46° 45' west. S. W. gale still blowing, weather continuing very pleasant. There passed close alongside, this day, two turtles, apparently of the hawk bill kind.

28th. The weather yet pleasant, the breeze now from the S. S. W. and freshening, the haze somewhat thicker. Latitude, at noon, 38° 44' north, longitude 38° 45' west. Variation, 15° 59' westerly. At 10 A. M. passed a ship standing to the westward. At 4 P. M. fell in with a mast, sent the boat to tow it alongside, hoisted it on deck to be made to answer any purpose that future emergency might require. While making the tow-line fast to it, many of the fish called by seamen leather-jackets, were swimming along by it; some of these were caught and cooked; they were very dry meated, and not good for food.

July 3d. Moderate breeze from the W. S. W. with pleasant weather. At 10 A. M. had sight of the island of St. Michael's bearing N. E. by compass about fourteen leagues distance. Our latitude, at noon, was 36° 56' north. St. Michael's now bore N. E. half N. about fifteen leagues distance. St. Mary's also in sight from the mast-head, bearing E. by S. apparently twenty leagues off: variation 21° 20'. Caught, with the grains, or barbed four-pronged iron spear, a fish called the Spanish mackerel: had too the good fortune to catch, with the boat, two turtles, one of the hawk-bill, the other a fine green chicken turtle; these last came very opportunely, and were welcome prizes.

5th. This day, had pleasant weather, with a light W. N. W. breeze, and a very smooth sea. At 6 P. M. were fired upon, and brought to, by the English frigate *Romulus,* from which ship an officer boarded the *Betsey,* and commenced examining her papers and men: having gotten through with this, he took leave, saying, as soon as we pleased, we might sail, and wished us a fortunate voyage. Latitude, at noon, 35° 27' north, longitude 25° 1' west.

12th. These twenty-four hours had the wind from N. E. attended with dark hazy weather. At 8 A. M. observed the sea water to be much colored, having all the appearance of water on soundings: on examining, by the help of a microscopic glass, into the cause, it was found to proceed from its being filled with millions of the *Neptunian animalcula,* a species of shrimp, Latitude, at noon, 21° 48' north, longitude 21° 51' west, variation 16° 35' westerly.

July 15th. Blowing a moderate gale from the N. E. with hazy weather. At 4 P. M. the look-out at the mast-head gave the welcome cry of "Land ho!" This was the isle of Sal, one of the Cape De Verde Islands, bearing W. S. W. distant about six leagues. The weather continuing thick and hazy, and finding it impossible to extend the view beyond a very limited distance, at 8 P. M. brought the brig to, and thought it most prudent to pass the night on short tacks, under easy sail. Passing a night in such a situation, keeps the mind in a state of painful uneasiness; many are the anxious wishes for day-break, which, when at last it does come, but little betters the prospect, unless the thickness of weather clears away; there is the continued order of "bout ship," and a dread that such will be followed by the cry of "land close aboard," or the more appalling fact of her striking against some rock which may not be laid down on the chart, or known to any on board; it is, in truth, a state of such anxiety as those only can have a correct idea of, who have been similarly situated. At 4 A. M. the thick haze having cleared off, bore away, and made sail. At 6 A. M. the island of Bonavista was in sight, bearing* S. W. half S. five or six leagues distance; the south point of the isle of Sal at the same time bearing N. W. four leagues distance. At 11 A. M. received a pilot on board, and at meridian were brought to an anchor in the harbor at Bonavista, called English road.

Dinner being got through with, went on shore, and paid my respects to the governor and commandant, (both offices being at this time vested in the same individual) by whom I was received very politely, and throughout the *Betsey's* stay, was treated with the most friendly attention. Arrangements were soon complied with him for the necessary supply of salt, as also some water, and refreshments.

---

*All bearings, unless they are particularly otherwise mentioned, are taken by compass.

While on the passage to the Cape De Verdes, by the expressed wish and counsel of the officers, it was thought advisable to alter the rig of the *Betsey*, and change her into a ship. This, it was supposed, and afterwards ascertained to be the fact, would be greatly to our advantage, for while laying off and on at the seal islands, to procure our cargo of fur seal skins, the cabin boy alone could tend and work a mizen-topsail, who certainly would be altogether unable to do any thing with the heavy boom of a brig's fore-and-aft mainsail. At this place the alteration was carried into effect: the mizen-mast, top, spars, rigging, sails, &c., were already in readiness, and the armorer, at his forge erected on shore, forged and made the chains, and all the other requisite ironwork, so that the mast was stepped, sails bent, and the *Betsey* rigged into a ship all ready for sea, in five days' time. This was accomplished without one dollar extra expense to the owners.

CHAPTER VII

PATAGONIA AND THE FALKLAND ISLANDS

THE Cape De Verde Islands have become so familiar to every reader, by the frequent reports given relative to them, that any farther description in this work is thereby rendered needless. We proceed, therefore, with the account of the voyage, by stating, that on July 23d, 1797, the *Betsey* having received on board a sufficiency of salt, together with a full supply of goats, pigs, fowls, fruit, &c., was got under way for sea; at noon, the S. E. end of Bonavista bore E. N. E. three leagues distance. The ship's course being S. E. by S. and steering for the South Seas, with a brisk trade wind from the N. E.

On the 30th, in latitude 10° 5′ north, longitude 19° 2′ west, the wind became variable, accompanied with very heavy rains, which continued falling, uninterruptedly, for the space of forty-eight hours, and until the wind settled between the S. S. W. and W. S. W. From this quarter it blew a gale, with but little cessation, until the 5th of August, when it veered around to the west, afterwards to the north, and finally to the E. N. E. where it remained for the greatest part of the 6th, when it fell calm. At 9 A. M. we had again a light breeze. This day we came up with, and spoke the ship *John and Elizabeth,* John Galway master, from Philadelphia, bound for Canton, out seventy days. At noon, our latitude was 5° 47′ north, longitude 14° 59' west, variation 15° 11' westerly: the wind now appearing to be settled into a moderate breeze from the S. S. W.; weather very fair, with flying clouds until the 10th. During this period, we caught, with hook and grains, as many of the Spanish mackerel, or bonetos, as were wished for; shoals of these fish, as well as the dolphin, being all around us; in fact, the men commenced the sport (as they called it) of taking them with the hook, and then plunging them back again into the sea. This piece of fun, however, was soon stopped. We caught also, with the harpoon, two herring-hogs, from among the great numbers that were around the ship, and tried out the oil they afforded, for the use of the lamps. On eating of the dolphin and mackerel, almost all on board were affected with a severe pain in the head, which shortly after was much inflamed; the eyes became red, and these distressing symptoms were attended with violent vomiting. Those who were thus affected, were evidently poisoned; the head and some of the limbs began also to swell, which swelling increased, until they had attained a most disagreeable form, having at the same time, a reddish cast over the head and limbs thus swollen. A dose, of from two to three tablespoonsful, of the sweet or olive oil, to each person thus affected, was found to give relief, but was not a preventive; for eating again, brought back the complaint. Whenever the fish, on being taken out of the water, was immediately cooked, and then eaten, no evil or unpleasant sensation was experienced; but, on the contrary, when first cured and dried, then cooked and eaten, the above poisonous effects appeared the more severe, and sooner in their operation, from such fish as had been dried by moonlight, and kept from the sun, than such as were dried wholly by the sun. This different effect, from the manner of preparing the fish, was plainly evident, on repeated trials of parts of the same fish on the same men.

August 17th. The wind veered and got settled from the S. E. by E. with fair weather and a pleasant gale. At 10 P. M. we crossed the Equator, in longitude 19° 50' west, our variation at this time being 11° 30' westerly. The Spanish mackerel and dolphin now began to be scarce, and by the time we had arrived in latitude 5° 00' south, they had entirely quit accompanying the ship; but previous to, and up to this day, the ship's company had had a daily and full supply of them.

Some men-of-war birds, boobies, and small white gulls, appeared around us, also numbers of those small but excellent pan-fish, the flying fish. These last mentioned were, while flying to avoid their enemies, or while sporting, nightly striking against the ship's sails, and falling thence on deck, in such considerable numbers, as that for several days the steward seldom omitted having a dish of them on the table for breakfast.

29th. We had a moderate breeze from the S. E. by E. with a smooth sea and pleasant weather. At 4 P. M. this day, the cabin boy, Henry, aged fourteen, while drawing a bucket of water at the fore-chains, fell overboard: the alarm was instantly given, and being on the quarter-deck at the time, and knowing that he was a good swimmer, I called out to him not to be frightened, but to rest and float on his back without fear, as a boat should soon be sent after him. He instantly turned as directed, and floated in this manner, until taken up some hundred yards astern. The ship at the time was going free from the wind, at the rate of four knots, and although the helm was immediately put a lee, and the main yard hove aback, yet it was some time before her way was deadened sufficiently to have the stern boat lowered away. By this time the little fellow was at so great a distance as not to be seen from the deck, and no one could have shown where he was, had it not been for the precaution taken of sending an officer, the moment the alarm was given, to the main top-gallant yard, to keep him in sight, and direct the boat to him; by which means he was soon discovered, laying perfectly composed. On being taken up into the boat, the officer asked him whether he was not tired or frightened, by waiting so long. He replied "No, sir! For as I passed by the stern, the captain told me to keep still, only to try to float, and not to be frightened, and that he would send a boat for me; so that I was not scared." This was uttered with the most innocent composure and cheerfulness. Our latitude at noon was 16° 12' south, longitude 31° 78' west, variation 00° 55' westerly.

September 13th. Lost the trades: the wind and weather now became variable. At 2 P. M. the land was to be seen from the mast-head. It appeared low and hilly; our depth of water at the time was twenty fathoms. Toward sunsetting, a large smoke was seen rising over the land; this was supposed to be a signal, by which the ship's appearance on their coast was made known. Latitude at noon, 30° 43' south, longitude 47° 52' west, variation 11° 15' easterly.

17th. Experienced a hard gale from the south-west, attended with thunder and sharp lightning, and very heavy rain, the ship laying to under a reefed mizen and mizen stay-sail. After being in this situation for sixteen hours, without receiving any damage, made sail again on her, the gale having begun to abate, and the wind veered round to N. by W. Our latitude at noon was 37° 21' south, longitude 52° 50' west.

The 22d, at noon, we were in fifty fathoms of water; saw some kelp, a few seals, and large numbers of birds, such as albatrosses, sheerwaters, sea or Port Egmont hens, gulls, and penguins. Latitude at this time, 39° 38' south, variation 17° 57' easterly.

28th. At noon, we had ground in twenty-nine fathoms; our latitude was 41° 20' south, variation 21° 19' easterly, with the wind light from the north-west, and pleasant weather; saw many seals, and large flocks of snipe.

October 2d. At 4 P. M. made the land, being part of the coast of Patagonia, bearing W. N. W. about seven leagues distant; a moderate breeze blowing at the time from the northward, with clear weather; our depth of water was at this moment, nineteen fathoms. Stood in for the land, and at 9 A. M. passed a shoal of breakers, one league or so distance from the shore; half an hour after, passed another shoal at a like distance from the shore, on which were easily to be seen a number of hair seals.

While running along the coast, at the distance of between three to five miles from the shore, we had soundings at the various depths of six, seven, and eight fathoms, the land appearing to be high, sandy or clay cliffs, with tableland back. We entered the bay of Sinfonda by the capes at its entrance, which are three or four miles apart, and sailed round and surveyed it without finding any harbor, or seeing any seals whatever, and were unable in any part, at a cable's length or two from the shore, to obtain bottom with forty-five fathoms of line. The land all around this extensive and singular bay appeared to be dry and barren, and not the least sign of fresh water was discovered. Latitude this day, at noon, 42° 25′ south.

We ran along this Patagonian coast as far as Cape Mattas, in latitude 45° 5′ south, but not finding the fur seals, for which we were in search, in any considerable numbers, left it, and steered for the Falkland Islands, where we arrived on the nineteenth, and came to an anchor in the harbor of Shallop Cove, at the cluster called the New Islands. On the passage from Cape Mattas, great numbers of whales and seals were seen.

So soon as the ship was safely moored, and dinner over, a boat was manned for the shore, among whose crew was Jonathan (of whom a slight description already has been given), who had obtained the promise, and was very solicitous to be with the first to land from the ship's company in this novel country. After sending on board the ship some geese and ducks which we had shot soon after landing, we started, taking along a large salt basket in which to bring back some eggs for supper; for the bird rookery, which lies up a valley, on the opposite or sea-board side of the island,

and is at a distance of something like three quarters of a mile across to it. This rookery (as it is called), contains, or extends over, a patch of ground of from four to six acres, on a side hill, surrounded with high bogs of the coarse grass called tussucks. Over this area, the birds, such as the albatrosses, penguins, and shags, have their nests, and to all appearance cover the entire surface in one grand assemblage. In fact, so closely side by side are they mixed together, that there is considerable difficulty in walking among them, great caution being necessarily exercised lest one should tread upon them, for they are so void of fear as to suffer themselves to be taken with the hand; yet in order to proceed, one is constantly obliged to push and kick them out of the way. On their part nothing backward, they return this rough manner of proceeding, with a continued pecking and biting at the hands and feet, frequently with such a painful nip as to start the blood. A continued cackling is kept up by this feathered fraternity to such a degree of clamorousness, that persons walking among them, within a yard of each other, cannot understand what their companions are speaking about, or even hear them, unless the speaker calls out in a loud voice what he has to say.

The albatross is the largest bird in this rookery. Their nests are built of mud, mixed with coarse grass, much in the form of a sugar loaf, but concave at the top, which forms the nest, about fourteen inches in diameter at the base, and from sixteen to twenty-four inches in height. On the top of this mound, its nest, sits this noble bird, in seeming pride and grandeur. They generally lay from two to four eggs, and in defense of these, or their young, will suffer themselves to be destroyed rather than abandon them. The seamen, in order to obtain their eggs, manage by the assistance of their seal clubs, to pry the bird off its nest with one hand, while with the other they gather the eggs; so soon as this is accomplished the bird resumes its former position, and soon lays more. Between each of these monuments for the nest of the albatross, formed of parcels or bunches of pebbles, mud, dry sticks, grass, and feathers, which they have been able to get together, are the nests (if they can be so called) of the penguin and the shag. These will also most stoutly defend their own, and a slap from the side-arms of the former, against the shins, is very painful.

The albatross begins to lay its eggs about the tenth of October; these are somewhat larger than those of a goose, having a shell of a dull white, the yelk being yellow, and if well cooked, makes a good dish for the table. The shag's eggs are speckled, with a blood-red yelk, and are not good for eating, having a strong fishy taste. The eggs most preferred of all that the South Sea country produces, are those of the Mackaronie penguin. This noble bird commences its laying during the first part of November: I have never known their eggs to be obtained at this rookery earlier than the second day of this month. These are a size larger than those of our domestic ducks, with a white shell, and much stronger than theirs; the substance being a little of the light blue cast, with a yellow yelk slightly tinged with crimson. They were always preferred by the officers; so much so, that while the ship lay here some were frequently served up at the cabin table with those of the common hen, cooked in different ways, and invariably selected on account of their superior flavor, and not being so dry as the hen's. There are four different kinds of this amphibious bird, viz. the King penguin, which is the largest, the Jackass penguin, the John penguin, and the Mackaronie: it is this last only, that inhabits the rookeries with the albatrosses; the other three keeping by themselves.

The Mackaronie is about sixteen inches high, and has on each side of its head a tuft of thin feathers, richly variegated in color, which gives the bird a very consequential and proud appearance. In its walk, or rather march, it is as erect as a soldier. One could sit for hours, and observe their manner of approaching the shore, after a spell of feeding in the sea; to effect this purpose, they make choice of a spot where the sea breaks directly against the side of the rocks, and while yet some seventy yards from the landing place, swimming moderately along in solid columns of hundreds together, toward it, commence diving and coming up again to the surface at short distances; this is continued until about within thirty feet of their landing, when they dive again, and come up in the surf ten or twelve feet from the rock, with such velocity as to land upon it perfectly erect, and clear of the surf; immediately forming in Indian file, and divided into distinct bodies, each division having its own leader, whom they follow, proceeding in their march up the valley or chasm, to the rookery, apparently with as much conceit of their superiority in point of discipline as ever a company of soldiers manifested on a public parade. The gratification derived from beholding a scene like this, is in a great measure counterbalanced, in the destruction committed among them by the sea-lions, which place themselves a few rods from the landing place, in the water, watching the time that the penguins are about to commence diving to land, at which period they are the most compact. At this moment, the lion settles himself under water with the intention of swimming under them, and when a suitable opportunity offers, rises suddenly in their midst, and seizes one or more of the birds in his jaws; then raising part of his ponderous body out of the water, he bites and shakes this, his prey, until they are torn in pieces, then devouring them. It frequently happens that some of these birds get badly wounded in the legs or wings, and land in this maimed condition; whenever this is the case, they are instantly attacked by their comrades, who peck and bite them until they rise up and take their places in the line of march, or until, by this tormenting, they are killed.

Of the existence and gentleness of such a large collection as this, his shipmates had often told Jonathan, on the passage; yet he was nevertheless far from being disposed to believe them, so that now, when the reality was to be unexpectedly presented to him, I had the curiosity to attend to, and see what the effect would be upon his mind. Keeping, therefore, a few rods in advance of him and his two shipmates, I sought a spot where the scene would at once be open to his sight, and for this purpose, selected a large tussuck bog, a little elevated, and near the edge of the rookery. Having advanced so near as to hear the birds, I beckoned to Jonathan to come to the same place, where, on raising himself by my side, and having the whole view before him of an extensive plain covered with this feathered race, while many were hovering over, and others sailing around the rookery, much like bees over and about their hive, nothing could equal his earnest and enraptured gaze: his eyes appeared to be ready to start from their sockets, the scene was so novel to him. "This, by 'gingoes,'* beats Vermont mountains all hollow;" he exclaimed, upon being asked what he thought of the birds in a South Sea rookery; giving his eyes no gentle rub with the hand, the better to assist their vision, he took another look, and turning suddenly around, he continued, "Oh! wonderful! who would ever 'a thought such sights as these were in the world!" "But this, Jonathan, you must recollect, is the other side of it, which, on shipping, you were so anxious to come and see." "Yes, yes, captain, so it is," said he, "and I am sufficiently well paid for coming, by the sight! Victuals and clothes are all I now want; you are welcome to my voyage, sir."

His two shipmates had by this time joined us, and the moment they saw Jonathan's singular appearance, they commenced, and continued for some little time, laughing in a most hearty manner. We now entered

*This was the utmost extent of swearing that he ever attained to during this voyage; and here I take the liberty to remark, that whenever in command of any vessel, it has been my constant rule, not to allow any swearing on board, and not to suffer an officer to curse, or use any low and vulgar language (to say the least of it), in my hearing, to the men; and nothing was ever lost by this rule, I am sure, by me. It was at all times with me a cardinal duty, to state, on shipping a crew, that it must be well understood by them, as a prominent part of our agreement, that all the quarrelling and swearing on board was to be done by myself, and the work by them. This has ever been readily agreed to, nor could I ever discover any advantage in governing seamen by coarse oaths, or swearing at all.

the rookeries, for the purpose of gathering some of the eggs, but Jonathan's strong desire to see yet more of what was so entirely new to him, was the cause of his neglecting all care of himself, and consequently of receiving some severe pecks on the cheeks, so much so as to draw blood, and while gathering the eggs, his eyes were really in danger from the beaks of the birds; however, by sundry frequent intimations of their nearness, they soon after taught him to be more careful, and in the present instance, towards the end, as it were, obliged him to assist in getting together and taking to the ship the day's collection.

I have kept these penguin's eggs in a good state of preservation, on board ship, for a period of nine months, by first immersing them in seal oil, though any will answer, then packing them in a cask with dry sand; a layer of sand, then a layer of eggs, and so on until the cask is filled, placing them all on their sides, with one end towards the bung, then heading the cask up, and stowing it bung up, in such a place as it can be got at on the third day, in order to be turned bung down, and so on; being turned every third day, until wanted for use, this method keeps the yelk from settling to the shell, and the sand mixing with the oil, forms a crust of sand and oil over it, by which the shell is kept perfectly air tight, and thus the egg is preserved from destruction.

On these islands and islets, which are said to number above three hundred, I have raised excellent potatoes, together with some tolerably good culinary vegetables, and salad; yet the soil, except here and there a small patch, is in general very poor, the surface being mostly covered over with a rank grass, growing in bunches or bogs, in many instances, over a bed of peat, which serves the purpose of fuel, while underneath is rock of different kinds, gray, blue, and slate colored. On these islands are to be found the horse, neat, or black cattle, and hogs, which have run wild on some of the large ones: on the English Malone, which is the largest of the group, is the wolf, or rather the wild dog, that Commodore Byron makes so much mention of. Several of the islands are also overrun with the white rabbits. The amphibious animals found upon them, are the sea-elephant, and the hair and fur seals. The waters around abound with several kinds of fish, such as different sorts of whales, the porpoise, the mullet, and others.

In almost every sandy bay or cove at these islands, in from two to six feet depth of water, at low tide, the round clam may be taken with the rake in great numbers, and as fine as can be procured in any country; there are also other shell-fish, such as the limpet and the muscle.

We found on the island the large white swan, with neck entirely black; the upland and lowland goose; the brant, a very beautiful and delicious bird; several kinds of the wild duck; the teal, an aquatic fowl, and the least of the duck species; divers; a variety of the plover; snipe; gulls; sea-hens; rooks, these last a most troublesome bird, being very mischievous and full of cunning; shags; albatrosses; the four different penguins, before mentioned; owls; and many smaller birds; some of them having plumage of the richest and most beautiful colors. Many of these can be taken at all seasons, and without much trouble.

Of berries, there is a great variety, and in the proper season, large quantities may be collected. These we ate either raw, or made into puddings. In fact, a person would be able to subsist at the Falkland Islands for a considerable length of time, without experiencing any great degree of suffering; for in addition to what has been specified, there is another article of food, and this is the root of the coarse tussuck grass, which when pulled up, breaks off close to its fibres; after taking off the outside cover, there remains a pulp about the size of ordinary sparrow-grass, which on being eaten, tastes very much like a green chestnut, and is very nourishing. By raising a dam across, near the mouth, of the rivulets and streams, and leaving a gateway that may be stopped at high tide, that excellent fish, the mullet, can be obtained in great abundance: they are equal, in my belief, to anything of the kind that the world can produce. For fuel, besides the peat already spoken of, there is considerable drift wood along the shores.

The water for shipping is of the best, and is to be had in an abundant and convenient supply. In the harbors, which are numerous and first rate, the tide rises from four to seven feet, at different places, and in some of the narrow passes, the tide is very rapid, which causes, over the rough or rocky ground, or sunken reefs, very dangerous rips.

The weather at these islands is most variable. Severe gales from the northward and south-west are of frequent occurrence, but at every season of the year the climate is healthy; seamen at all times relish their meals with the heartiest appetite, and with the exception of slight colds it is very seldom any of them complain of sickness. The temperature of the climate is so even, that in summer the heat never puts the men to any inconvenience, and in winter, although the snow storms are very frequent, yet it seldom remains on the ground, unless at the tops of the mountains, over forty-eight hours at any one time; and rarely is the cold so severe as to freeze the surface of a small pond sufficiently hard to admit of a person's crossing on the ice.

From the information obtained of Captain O. Paddock, master of the whale-ship *Olive Branch*, of Nantucket, which stopped at these islands for water and refreshments, while the *Betsey* lay here, I was induced to believe we could at times safely land, and take dry skins at the island of Massafuero, and notwithstanding this was in direct opposition to my previous advices, yet from certain former transactions with Captain Paddock, I was confident that the utmost dependence could be placed upon his word: he also mentioned his having more than once stepped from his boat on shore, at that island, without even wetting his feet, when the wind was from a particular point, and satisfactory confirmed the intelligence previously received, that the fur seals were numerous at that island. From this information, and the fact of our not having been very successful here, I thought it best to visit Massafuero. A goodly number of geese and other refreshments, among which were fifty-six barrels of the favorite penguin eggs, together with plenty of water, were accordingly snugly stowed on board, as sea stock, and Jonathan, who had become a great favorite with all on board, and to whom his shipmates, since our departure from the United States, had often given an ounce out of their allowance, from good feeling; being a great eater, as well as a wonderful hand at hard work, was allowed the extra privilege to fill two tierces with these eggs for himself; this he did when off duty, his shipmates cheerfully assisting him in his collection of them. From having made use of dried kelp as fuel, while at these islands, and which in its dry state was found to burn well, we were induced to take on board a supply, as sea stock; but found, to our sad disappointment, after being a few days at sea, that it became moist and soft, and when in this state, would not burn to any advantage, even when mixed with wood, so that had we depended altogether on the dried kelp as our fuel, for cooking purposes at sea, we should have been led into a very unpleasant and uncomfortable situation: it is, however, good fuel on shore, when used in its dry state; it then burns freely, and makes a good coal.

CHAPTER VIII

PASSAGE ROUND CAPE HORN

DECEMBER 8th, 1797. Having made every necesesary preparation for sea, at 9 A. M. got the ship under way, and stood out through the south passage (having in company the *Olive Branch,* Captain Paddock), with a fine breeze from the west, with passing clouds. At meridian, the New Islands bore N. N. E. six leagues distant. Our course was now to the S. S. W. At 2 P. M. we had laid the land of these islands below our horizon; the ship, the while, surrounded by a variety of oceanic birds.

On the 11th, had the wind light from the N. W. with clear weather. At noon, latitude 54° 12′ south; had sight of the island of Staten Land, bearing S. S. W. about twelve leagues distant. In the evening the wind began to fail, the little there was keeping continually changing, the sea at the time very smooth.

While walking the quarter-deck this evening (an exercise I usually took at this portion of the twenty-four hours), my man Jonathan came aft, and after folding his arms together, and for a while swinging his hat in one hand, backwards and forwards, with all the regularity of a pendulum, his body somewhat bending forward, a position he usually assumed when desirous of obtaining any favor, he began the following short dialogue, in which he made known his wants and desires, by saying, "If the captain is willing, I have a favor to ask."

"Well, and what is it, Jonathan?"

"Why sir, Ben and Tom says, that the land we saw before it got dark is an island aside the Patagonian coast, where these Patagonian giants, of whom folks talk so much about, live; and as they expect the captain is going to anchor in a harbor there, we have been making up a plan by which to catch one of them giants, so that by the time we get home, we can have him tamed, and then I can show him about the country, and making a swinging great deal of money by it."

"Well, but my dear fellow, these gigantic Indians are noted warriors, and as it is reported, a very wild race; they are also said to be very large and strong; now how can you expect to catch one of them?"

"Yes, sir, Ben and Tom both have said that they were real giants, that their heads are so big that some of them measures more than a foot between their eyes; and that's what I so dreadfully want one for, he'll be such a sight to show about."

"Well; now, if they are so large, they must most likely have great strength in proportion; and how can you manage so as to master and catch one?"

"By gosh! captain (the immense wealth his imagination had pictured out, as the grand result of exhibiting one of these people in America, completely setting aside all doubts as to a successful result of the undertaking), I guess, I think I know how to do it easy enough too; and that's just in the same way as we catch bears among the mountains in Vermont."

"Well, how is that? How do they catch bears?"

"We take a large tub, sawed from the head of a hogshead, and place it where the bears come; then we put into it two gallons of molasses, and one of New England rum, and a little water, and then stir all up together; very soon after we get out of the way, the bears come along, and drink, and keep drinking, till they get as drunk as a sow, and then we can do any thing with them we please. Now that's the very favor I wants of the captain when the ship gets into the harbor; to have the tub, the molasses, and the rum, and as I gave up my voyage to the captain, while at the rookery, Ben and Tom says they will have the things charged to them."

"And if you catch one in this way, by first making him drunk, what will you do, Jonathan, when he becomes sober again? Is there no danger, when he shall find out how he has been trapped, that in his rage, he may throw all hands overboard, especially, if they are so large and strong as you tell of?"

"Oh, no sir, we have fixed all that too; for while he is sleeping soundly, after drinking from the tub, we are going to lash him down to the ring bolts, on deck, and there keep him bound until he gets tame."

"Well then, you shall have all these articles, Jonathan, if we anchor in their harbor; but this is very doubtful, however, for if the wind should be favorable, time will not suffer it."

This was said to ease his mind, for some promise like it I found to be necessary, as he was bent upon accomplishing this scheme, which his shipmates had no doubt put into his head.

The wind having by this time settled into a moderate gale from the N. N. W. we crowded sail, in order to pass Cape St. John's, the most eastern promontory of Staten Land, off which we observed a strong current, and very dangerous rip.

12th. The gale from the N. N. W. still blowing, at 2 P. M. we passed Cape St. John's, latitude at noon, 55° 8' south, St. John's then bearing north by west, seven leagues distant.

On the 13th, the wind veered around to the westward, and at 6 P. M. suddenly burst upon us in a heavy squall from the S. S. W. in consequence of which we were obliged to heave the ship to, under a reefed fore-course and storm stay-sail. In the heavy sea running at the time, I had the satisfaction to observe that our little bark, to use the seamen's phrase, "behaved as lively as a duck." At 8 A. M. the gale had abated, yet there was a continued succession of snow squalls after this during the day. At noon, our latitude was 55° 30' south, the high land of Cape St. John's still in sight, bearing N. W. by W. about thirteen leagues distant.

Up to the nineteenth, nothing occurred worth noticing, except that we had continually variable and violent gales, attended with severe squalls of snow and sleet, and a heavy sea, our little ship still working admirably; not, however, when she came in contact with the large seas, without giving all on deck, even those quite aft, a wet jacket; the usual salute in doubling Cape Horn. Our latitude, at noon, was 60° 31' south, longitude 69° 37' west, variation 28° 20' easterly.

I would here note the fact, which has been proved

in the recent experience of commanders of our South Sea sealing vessels, on the passage round this noted cape into the South Pacific, since the re-discovery of New South Shetlands, and of the continent of Palmer's Land, to the south of them, that a ship desirous to make the passage, the most favorable and advantageous in every point of view, should, on leaving Cape St. John's, stand boldly off to the south, until she arrives up to the latitude of 63° 00′ south, keeping constantly with the prevailing southwesterly gales, on her starboard tack; by so doing, should these gales flat her off to the eastward, so as to fall in with the South Shetland Island, while endeavoring to obtain the above latitude, no disadvantage will occur, for it will not be long after getting sight of these islands, before the wind will be round from the southeastward, when by keeping between the latitude of 63° and 65° south, she will have easterly winds, so as to be enabled to run up her latitude in a short time. A ship, by taking this track, in doubling Cape Horn, will arrive in the South Pacific with far less injury to her hull, spars, sails, and rigging, and what is certainly a weighty consideration, and very essential to their health, with less drenching, hardship, and fatigue to her crew, than by taking any other; besides making the passage by this route in a shorter time.

Our sealing vessels have invariably made their passages from the South Shetlands to the Island of St. Mary's, on the west coast of Chili, in from fifteen to twenty days, by running well to the westward, between those latitudes, and then, when they come to bear up to the northward, and receive the westerly gales as they advance north, which blow almost continually on the Cape Horn side, as well as in the middle of the passage between the South Shetlands and this cape, they run their northing up with a free wind, and make a quick passage of it. This track is therefore recommended in doubling Cape Horn for the Pacific, at all seasons of the year; as it is well ascertained, that the west and south-west gales, on the cape side and in the stream, effected (as is supposed) by the mountains, of Palmer's Land and the islands, blow at the same time from the south and south-east, above the latitude of 63° south.

On the 24th, having had a calm for several hours, and, for Cape Horn, a smooth sea, we lowered away a boat, for the purpose of ascertaining the current, which was found to be setting to the north-east, at the rate of three quarters of a mile per hour: our ship still accompanied by many penguins, and numbers of oceanic birds; we took advantage of this and the calm, and shot as many sheerwaters, &c., as were enough to make a pot-pie for all hands.

On the 29th, the calm of the 24th was succeeded by a light breeze from the north, which soon increased to a gale, attended with rain; were obliged to bring the ship under reefed courses and storm stay-sails. At 11 A. M. it began to moderate, the clouds broke away, and the wind shifted to the W. S. W. At noon, our latitude was 54° 58' south, longitude 76° 20' west. At this time had sight of the land of the west coast; a high cape, bearing north, about twelve leagues distant, and although it was midsummer here, the weather soon became so squally, with snow and sleet, as again to close the sight of land from our view.

January 1st, 1798. Winds light from W. S. W. and at intervals during the day, squally, with rain. At 6 P. M. saw the land again, bearing N. E. at a very great distance, appearing to be very high. Latitude, at noon, 53° 55′ south.

6th. The wind, for the several days past, has been varying between S. and N. W.; at times moderate, and then squally, accompanied with a great deal of rain. At 7 A. M. it suddenly burst upon us in a squall from the north, with rain; and at 11 A. M. we had it, a gale from N. by W. and cloudy: saw the land again; tacked ship, therefore, and stood off shore. The birds, such as albatrosses of two or three kinds, black and gray sheerwaters, also large flights of snipe, were around the ship continually, as well as numbers of small land birds, which came on board, lighting upon the spars and rigging, and wherever they could find a resting place. The sea water was this day observed to be highly colored. Latitude, at noon, 46° 45′ south, longitude 80° 56′ west, variation 18° 33′ easterly.

13th. The weather clear and calm, with a smooth sea; made trial with the boat to ascertain whether there was any current, but found none. Saw several shoals of sperm whales, and many seals. Our latitude, at noon, was 34° 33′ south, longitude 80° 10′ west, variation 11° 24 ′easterly. Had now a light breeze from S. S. E. with pleasant weather.

19th. The wind, during the last five days, remained between the S. S. E. and S. S. W. points of the compass, with a moderate breeze, the weather still continuing pleasant. At 6 A. M. had sight of the island of Massafuero, bearing W. half S.; at 7 A. M. the Island of Juan Fernandez was also in sight, bearing E. by N. Hauled up for Massafuero, and at 10 A. M. was near its northern shore, on which, with the assistance of the glass, numbers of seal were seen.

CHAPTER IX

MASSAFUERO TO THE MARQUESAS ISLANDS

JANUARY 20th, 1798. In order to gratify the officers, who were very anxious to have a nearer view of the seals, as well as to reconnoitre the landing, the ship was hove to, and two of the boats hoisted out for the above purpose, well manned and supplied with every article that would be requisite. As they put off, the officers received particular orders not to let their anxiety overrule their better judgment, so much so as would induce them to endeavor to effect a landing, if the surf, which from the ship was to all appearance very high, should, on nearing it be found to be dangerous. I was the more urgent on this point, as previous to what Captain Paddock had stated; that as often as every fifth or seventh day, there would be no danger in the attempt; that then one could step from the boat on shore without even so much as wetting the feet; and which information was subsequently found to be correct, I had been induced to believe, from various reports, that it was not practicable to make a landing on this island without swimming through the surf; consequently no dry fur seal skins could be brought off without their getting wet before reaching the ship; it would therefore be useless to come here for them. Under this impression, when the boats put off, I was very desirous that no attempt of the kind, viz. at a landing, should be made, lest something might occur which would have a tendency to depress the spirits of the men, and discourage them from making any farther exertions, but would rather have waited until such days when there should be no surf, in the meanwhile laying off and on with the ship. As the boats neared the shore (being yet outside the outer rollers) on it, together with the north-west plain of the island, between three and four hundred thousand of these fur seals were at once presented to their view: this was a temptation so powerful, as to induce the second officer, regardless of all orders and instructions, and the present advice and consent of his senior officer, to determine upon immediately trying to land.

His boat's crew were, as well as himself, all good swimmers, and counting upon this as their security in case of an upset, until their comrades in the other boat should take them up, or they be able to reach the shore in safety, they obstinately and imprudently decided to pull boldly through the surf, and effect the landing on the rocky beach at all hazards. When within one or two hundred yards of the desired point, the boat was struck by an over-sized breaker, and capsized, the bow striking against a rock under water, and breaking her into two parts in the first blow; the men, by swimming and diving, and managing the best way they could, were at length enabled by God's Providence, to reach the land, every man maimed or bruised in a greater or less degree, by the violence with which the surf had thrown them against the rocks, losing every article they had except those on their bodies; as for the boat, little of it was to be found, for literally speaking, it was stove into more than a hundred pieces.

VIEW OF MASAFUERO, NEAR JUAN FERNANDEZ

From an engraving in Anson's *Voyage Round the World*. London, 1748

VIEW AT LA CHRISTIANA, MARQUESAS ISLANDS

From an engraving in Shillibeer's *The Briton's Voyage*, Taunton, 1817

After waiting until he had seen the men one by one safely reach the shore, the senior officer returned to the ship with this sad report. It was now blowing a gale; the boat was therefore hoisted in on deck, after which tacked ship off shore, and trimmed sails so as to keep to the windward, and hold on to the island; all hands were then mustered aft, and informed of the precipitate act of the officer on shore, and the consequent misfortune resulting from it, yet notwithstanding this, there was no cause to doubt our, eventually, succeeding in procuring a cargo for the China market at this island; moreover, it was to be hoped that no one would suffer himself to be in the least discouraged, adding the assurance that had been received from Captain Paddock, of the possibility of landing at particular days, and which there was no reason to call in question. The steward, at the conclusion, received orders to "splice the main brace," and the men giving three cheers, repaired to their several duties. Next day it blew a severe gale, and was so squally as not to permit even our indulging a hope to land; but on the second day it moderated, and in the after part of the same, while standing in shore as near as prudence would allow, we observed by a signal affixed to an oar that the men on shore were in want, hove the ship to, therefore, and hoisted out a boat, into which were put a strong iron bound sixty-gallon cask of water, and another of bread. An active seaman, as well as good swimmer, who was applied to, to guide these two casks from the boat at the grapnels through the surf to his shipmates, by swimming one at a time in such a direction as to cause them to land at one of the smoothest places clear of the rocks, answered he thought he could, and was willing to do his best.

The boat then put off, and soon after returned with the information that both the casks had been safely guided to our men, who by signs had made it known, that they had suffered much for want of water. The evening now setting in, we hoisted the boat on deck, and trimmed ship to keep to the windward, intending to make trial again with her the following morning, as the wind and present appearances encouraged us to hope we should then meet with more success. Accordingly, early in the day, stood close in with the ship; the surf, as now seen with a glass, did not appear to be so high as the few days past.

In addition to a similar quantity of water and bread, there was put into the boat two coils of deep sea line, each one hundred and twenty fathoms in length, together with sealing apparatus, all of which were to be landed, if the surf would allow the boat to approach within two hundred fathoms of the stony beach, so that the line might be made to reach it. By the bright countenances of the crew, who shortly after returned, we were prepared to hear some good news. And so it was; for they reported the complete success of the past exertions, and that on the first trial by swimming with the end of our line to the shore, the seaman had guided the casks in safety, and that they had disembarked every article in the same condition; likewise stating that the surf was now become so moderate as to admit of similar operations. This last was confirmed upon the subsequent return of the boat, when it appeared that the swell had so far subsided as to admit of their resting the bows on the beach, and thus land the articles from her.

Upon this, our communications with the shore were pushed on with additional activity; so that by the middle of the afternoon, every thing necessary was landed, and the first officer having gone on shore with a fresh gang of men, had sent off the second, together with all the bruised invalids, to the ship. Our business of procuring a cargo of fur seal skins was now commenced, and the practicability of continuing it, and embarking the skins safe and dry on ship-board, established beyond a doubt.

From the statement of the second officer, it appeared that they had not been able to discover any fresh water on this part of the island; and in consequence of this, and not finding any pass over the mountain to the other parts, the sufferings of himself and shipmates had been very severe, until their receiving the first cask which had been sent off to them. As to food, although almost entirely destitute of any, as well as fire, or means by which to make one, yet they had managed to get along tolerably well, by cutting the meat of the seals into very thin slices, and laying these upon the rocks, exposed to the sun until perfectly dry. They also caught a few fish, which were prepared in a similar manner, and made to answer as provisions. Nor were we able, during our stay, to obtain any fresh water, on the north-west side of the island; but upon strictly searching, we discovered a passage over to the other parts, and up the east side of the mountain, some good water; also near its base on the east side of its south-east point, a spring of excellent water was found, with a deep pool, or basin, some twenty by thirty feet in circumference, attached to it, abreast of which is a small beach, very convenient for shipping to fill their water, by anchoring eighteen or twenty fathoms off; while at a short distance, fire wood can be obtained.

While we remained here, the weather for the most of the time was pleasant, the wind generally blowing from the S. S. E. and S. W. with once in every two or three weeks a gale from the north, which brought along with it plenty of rain, invariably succeeded by torrents of water, rushing down the gullies, by this means greatly increased in size, bringing along stones, rocks, earth, and trees, which last the violence of the gale had torn from their roots, and on their way down the mountain, were broken in pieces and left in heaps in the niches on the sides of the gullies, where the wood becomes dry and well seasoned, and in fine condition for ship use; it is also very convenient, and easily to be procured. In the woods, on the mountain, there are many kinds of small birds; the goats we found in great numbers, and gentle, or rather, not very shy; it was no very difficult matter to walk around and huddle them closely together and in this manner, with a single discharge of buckshot, kill from three to five; they were found to be fat, sweet, and excellent.

The waters of Massafuero abound with a great variety of the finest fish, among which are the halibut, cod, sea bass, &c., &c. The large grouper, caught here in deep water, forms the chief part of a dish that would satisfy the desire of a most fastidious epicure. A supply of sea bass, and other fish, was obtained by the men in a manner seldom practised, but not the less successful on that account: thus, while washing out the seal skins at the edge of the water, bits and scraps of meat and blubber were thrown into it; these the fish would most eagerly seek to get hold of, and shortly became so ravenous for them as to come up with their heads out of water, near to the seamen, and thus exposed, were easily killed with the seal clubs.

The only places for anchorage at this island are on the east side, abreast of the watering-place, and off the north-west side, abreast of the plain; at this last mentioned place, we frequently anchored in twelve to fifteen fathoms, over a bed of rocks and sand, for the purpose of receiving our cargo.

By the second day of April, we had procured, and conveyed safely on board, a full cargo of selected fur seal skins; indeed, so anxious were the officers and men to make sure of filling the ship, that even after the hold was stowed so as not to have room for any more, then the cabin, and finally the forecastle, were filled, leaving just space enough for the accommodation of the ship's company; and yet there was remaining in stacks on shore, more than four thousand skins; with these, a boatswain and boat's company were left, to take charge of, and add to, until a vessel from our owners should call for them. These dry skins, after being stowed on board ship for a few days, in tiers, will settle very much. This was the case with our own; in the course of a couple of weeks we were enabled to clear the men's abode in the forecastle, and in like manner a portion of the cabin; but notwithstanding this, leaving still on deck a small part of our bread and salt provisions, with at least the full half of our wood and water, for some length of time afterwards.

Every thing was now actively preparing for sea; wood and water, together with an abundant sea stock of goat's meat, was got on board, and carefully secured. Thus, in the space of ten weeks, by perseverance and industry, was our little ship completely laden to cross the Pacific to Canton, for a market; every one in fine spirits for the performance of farther duties. At the time of leaving Massafuero, there was, according to our computation, between five and seven hundred thousand fur seals there, and several thousand wild goats; subsequently, there has been but little short of a million of those fur seal skins taken at this island, nearly all of which were carried to Canton, and there exchanged for China goods, suitable for the home market, which must have paid several hundred thousand dollars into the treasury of the United States. Such an amount from this isolated spot, is one evidence in the many of the important advantage American enterprise, in this fishery and commercial trade of the Pacifics, has heretofore been to the nation; therefore, the obligation on government by exploring, to discover new places and sources for its continued support.

April 5th, 1798. We left Massafuero at 6 P. M. and stood to the N. W. for the purpose of getting hold of the true trade wind, and then to shape our course for the Marquesas Islands. At a distance, as Massafuero sunk below the horizon, the portion of mountain (about three quarters) yet in sight, appeared in form very similar to that of a shoe.

8th. The island bore S. S. E. four leagues distant.

10th. Pleasant weather, with a fresh breeze from the S. E. Latitude, at noon, 27° 28' south, longitude, 80° 23' west. In the evening, large flights of birds passed by the ship, making a noise precisely like that made by land-snipe.

15th. The weather still continues pleasant, with a moderate trade wind from E. S. E. Latitude, at noon, 22° 16' south, longitude 86° 43', west. Unlike the course pursued by able and celebrated European navigators, as appeared in the various accounts of their voyages which we had on board, wherein it is asserted, that they were under the necessity of putting into port, even in those trade wind latitudes of the Pacific Ocean, in order to refit and repair their ship's rigging, &c., we accomplished the operation of stripping a ship's masts, and putting on an entire new suit of rigging, at sea. As nothing of the kind had ever been performed before this, or if it had, never to our knowledge, it must of course be considered as an American precedent.

Previous to the ship's departure from New York, a sufficient quantity of rope, for a new gang of standing rigging, had been taken on board; this had been prepared, during our leisure hours heretofore, and fitted, in readiness to go over the mast-head when necessary. For several days past, the trade wind had not veered more than two points, blowing directly on our stern, so that all the sails which would be of service, could be set on one mast. This was a most favorable opportunity for replacing our old rigging with the new; we, therefore, hit upon the expedient of stripping one mast at a time, and accordingly commenced with the foremast, securing it well by purchases and tackles at the hounds, before letting up the lower rigging, afterwards putting the new suit on the lower mast, topmast, &c., and then setting all taught up, preparatory to spreading sail on it; as soon as every thing was completely ready, this was done, and the main-mast served in a like manner. (It will of course be recollected, that the mizen-mast and rigging were new when we left the Cape De Verde Islands). Our carpenter and armorer were first rate workmen, and had made a set of machinery, by means of which from the old rigging we layed a new set of running rigging throughout; thus equipped, our ship, as respected her rigging, was now in a far better condition than when she left the United States. This was a great consolation, the more so from the probability, that at the time we should enter the China Sea, we might expect to meet bad weather, and perhaps a typhoon.

20th. Our fair wind still holds on from E. S. E. as well as the agreeable weather so long enjoyed. We had this day in company, some tropic and egg birds, as well as our little Neptunian favorites, the flying fish. During the few nights past, the dew has been very heavy, falling from the sails, as after a rain; we likewise experienced a swell from the southward. Latitude, at noon, 17° 38' south, longitude 95° 53' west, variation 5° 4' easterly.

27th. Wind and weather, same as for the few past days. Several shoals of herring-hogs passed around and by the ship, as well as great numbers of oceanic and land birds, such as cape hens, a solitary small hawk, egg birds, and others. Latitude, at noon, 14°

18' south, longitude 106° 35' west, variation 3° 39' easterly. Our ship now under full sail, steering-sails on both sides, alow and aloft, and sky-scrapers. The nights now are, and have been for some time past, most delightful, with the atmosphere overhead so clear, that the stars appear to shine with additional brilliancy. It was not until May 2d, that the weather began to show any signs of alteration; when, for the following twenty-four hours, we had some squalls of rain, with thunder and lightning. At 6 P. M. had it calm, with a smooth sea; embraced the opportunity, and made trial to discover the current, but without success. Latitude, at noon, 10° 30' south, longitude 116° 33' west, variation 2° 43' easterly. The little white gulls hovering over the ship, together with some small land birds, evince a disposition to light. Had once more in company, the Spanish mackerel, flying-fish, and now and then a dolphin.

17th. Had the trade wind from about east; at times, a passing cloud, from which fell light showers of rain. Latitude, at noon, 9° 25' south, longitude 137° 28' west, variation 2° 5' easterly. Saw this day several shoals of sperm whales, black fish, and herring-hogs.

19th. A light trade wind from the same quarter as on the 17th, with hazy weather. At half past three P. M. the man on the lookout, at the top-mast-head, gave the welcome cry of "Land ho!" bearing W. by S. about six leagues distant, stating it to be a high round island, in form of a sugar loaf; this, on a nearer view, proved to be Hood's Island. At the distance of forty-eight leagues to the eastward of this island, we began to fall in with shoals of sperm whales, and their frequent neighbors, the round blunt-nosed fish, called by the whalemen, black fish; these increased in numbers, until we had arrived within three leagues of this island. At 5 P. M. the Island of La Domineaque was in sight, bearing S. W. by W. about seven leagues distant. At 8 P. M. brought the ship upon a wind; passed the night under easy sail, on short tacks, to keep our station to windward of the land. In the morning, at five o'clock, bore up, and made sail to the westward, towards the land; an hour afterwards, had sight of the Island of St. Pedrie, bearing S. S. W. five leagues distant. This island is said to be inhabited by the handsomest built race of people of all the South Pacific Islands. At 8 A. M. the Island of La Christiana was in sight, bearing S. W. by W. half W. seven leagues off. Wore ship with the intention of running down along the north coast of La Domineaque: as we came abreast of the valleys, the green foliage of the trees presented a most enchanting prospect to our eyes, having been so long estranged therefrom. At 11 A. M. a double canoe came alongside, in which were eleven of the natives; on heaving the ship to, some of these clambered up the side as far as the plank sheer, on which they remained, holding on by the rail-way; nothing could induce them to come over on the deck; they nevertheless very readily bartered their cocoanuts, &c., for nails and small pieces of iron hoops. This traffic being soon finished, we bore away again to the westward, alongshore, at noon being near the middle of the island, and but a short distance from it; our latitude was 9° 40′ south.

CHAPTER X

THE MARQUESAS AND WASHINGTON ISLANDS

MAY 20th, 1798. Had pleasant weather, with a moderate breeze. At 2 P. M. hove the ship to, abreast a valley covered with the bread-fruit and cocoa-nut trees. We were soon visited by three canoes, from a village near by; the natives bringing with them, however, no other articles for trade than a few eels, and some squid, a small fish, of a sweetish and rather unpleasant taste; these we thought proper to decline purchasing, and filled away again. At 5 P. M. brought the ship to again, opposite another village, from which several canoes put off for us, yet there were but two of these that we could in any way persuaded to come alongside; from the natives in these, we were soon made to understand the cause of their fear; for the first sign given by them was to inform us that they were afraid of our carriage-guns. This source of disquiet was therefore speedily removed, by causing all the guns to be run in, and the ports to be shut. Their companions observing this last movement, immediately came alongside; not yet however, entirely divested of dread, or in full confidence, yet notwithstanding this, we managed matters so successfully as to obtain from them a few cocoanuts, and a sort of pudding, made from the breadfruit, also some of their fishing lines in exchange for small pieces of iron hoop of from four to five inches in length, and some nails. Iron, in any shape, they were most anxious to obtain; the beads and toys were held in little estimation; the small looking-glasses, and bright buttons, when handed to them they would turn over and over, examining every part very carefully before they gave up their articles, then after pondering the *pros* and *cons,* they would return the glasses, and point to the pieces of iron hoop.

At sunset, our visitors had taken their leave, and when clear of us, we hauled off shore, under easy sail, upon a wind, for the purpose of keeping our station during the night. The fragrance of these green valleys, brought off to us by flaws of wind at intervals, was truly delicious, and a person that has at no time enjoyed it, can scarcely be able to conceive with what delight we received it, after having been for a length of time at sea; it actually seems to take hold upon the feelings in such a manner as to reanimate the whole system.

At day-break, stood in for the shore. At 8 A. M. brought the ship to, abreast of a village, from which in a very short time we were visited by a number of the natives, in their canoes, bringing along articles for barter. Our guns having been previously run in, and the ports closed, no difficulty was experienced in inducing them to come alongside; a brisk trade was therefore speedily commenced, bits of iron hoop, nails, and knives, being given in payment for cocoa-nuts, bread-fruit, fish, and fish-lines. After trading until all they had brought off was sold, they left us, and we bore up, and at noon the north-west end of the island bore S.

S. E. two leagues distant, at which time we hauled on a wind for the Island of La Christiana.

On the 21st, we were employed in working to windward. During the night had heavy showers of rain, and squally weather. At 8 A. M. when close up under the Island of La Christiana, and near to Cook's harbor, into which we purposed working the ship, for an anchorage, as well as to obtain water and refreshments; several canoes came off, containing a greater number of the natives than we had yet seen together. This induced me to heave to, and an active trade was again established. As was the case with the other natives, so here; nothing but cocoa-nuts, bread-fruit, and small fish, could be had, and for these they wanted powder, knives, tools, axes, hatchets, and the like. The first mentioned article I refused giving them entirely; in hopes too of being able to persuade them to bring off some hogs, &c. the axes and hatchets were held back. It was in vain, however, and I was led to believe that these, with fowls, and other productions, were not very plenteous among them; this was subsequently found to be the fact. At this moment, two personages, who, from the great deference shown them by the others, their superior equipments, and the large number of attendants (there being some thirty or so) were evidently men of rank and influence, came alongside in a large double canoe; yet even these had none of those articles we were so anxious to procure. We were soon given to understand, that their wishes were, to have the ship brought into the harbor; promising also, when this was done, that we should be supplied with hogs, and all things, in great abundance.

From the friendly bearing of these people, I was induced to believe we might be greatly benefitted in having their assistance in piloting the ship into the harbor; but after remaining on board some hours, during which time every exertion had been made, by working to windward, to obtain the desired point, without success, they left us a little past noon, exhibiting while on board, every appearance of discontentedness, and an impatience to be getting on shore; nevertheless, as they pushed off, making us many friendly and inviting signs to come soon to an anchor in their harbor. The double war canoes had each on its bows, by way of ornament, four human skulls, and as we were examining these, the chiefs offered to part with them in barter; but not being the sort of refreshment for which we were seeking, their proposal in this case was refused. While attempting to get in, the wind came off in such heavy gusts and whirls, that we were frequently obliged to clew all down and up, so that instead of gaining ground, we lost.

22d. In addition to the hard squalls and gusts of wind of the previous day, we began to have at 1 P. M. heavy showers of rain. This was a trifle too uncomfortable for our Indian acquaintances, who were not long in making up their minds as to quitting. At 3 P. M. on the lighting up of a rain squall, a small canoe, in which only two persons were to be seen, was observed hastening towards the ship, coming from the western part of the island, or from some spot to the westward of the harbor. All the natives had recently left us, and who this stranger could be, was a question none on board could answer. It appeared to be so singular a circumstance, that notwithstanding the imperative necessity there existed for securing an anchorage in the harbor, if possible, before night should close in upon us, that the ship was hove to, to wait until these persons should arrive something nearer. As their small canoe came alongside, we were greatly astonished to hear one of the persons exclaim in our mother tongue, "Sir, I am an Englishman, and now call upon, as I have come to you, to preserve my life." Words cannot express my surprise at this moment, on hearing so unexpected a claim. The stranger was instantly assisted in getting up the gangway, and no sooner had attained the deck, than observing, "I am a missionary," he sank into a seat provided for him on the quarter-deck, and bowed his head for a few minutes, in this position returning thanks to that Heavenly Being who protects even the sparrow; meanwhile, regardless of those around, he seemed only anxious to acknowledge his Creator's kindness in thus once more giving him freedom. After receiving the assurance of being among Christian friends, and becoming a little more composed, he arose, and proceeded to give an account of past transactions on the island.

"Thank Heaven! sir," I answered to one of his inquiries, "you are safe." He then stated himself to be the Rev. William Pascoe Crook, whom the Missionary Society in London had sent out to these islands, where he had been landed some months past, from the missionary ship *Duff*, Captain Wilson; that the recent, as well as the present disposition of the natives towards him, had kept his mind in a continued state of uneasiness for weeks past; that in two instances of narrow escape, he owed the preservation of his life, under God's blessing, to his friend, the native chief who had accompanied him on board, and whom he at this time introduced, adding the wish to remain with the ship until he could be landed in some place of safety. In reply, I observed, that the character he bore was a sufficient recommendation to insure for himself all the comforts and accommodation our ship could afford, and that he was at liberty to consider her as his home, and make use of the cabin as freely and equally with myself, until we should arrive at New York again.

After introducing Mr. Crook to the officers, and requesting their particular attention in his behalf, together with his friend the chief, he was led below, into the cabin, where, upon being seated, my limited wardrobe was spread before him, with a request that he would select for himself. Mr. Crook was at this time dressed in the native garb of the island, having only the maro on (a piece of cloth manufactured by the natives, which wound around the middle of the body, with one end passing down in front, is tucked up at the back, under the part which goes around the body); the remaining portion of his person, from being continually exposed to the sun, had become tanned nearly as brown as the chiefs themselves were; and this mode of dress he had been under the necessity of submitting to for months past. At his request (he thinking it would not be judicious to choose out or accept any portions of dress so long as his friend the chief remained on board), the selection of garments was left until the chief should go on shore. At the same time Mr. Crook stated that he felt very anxious to communicate to me some information respecting the state of the island, which would have a reference to the government of my future proceedings; as he conceived, from the knowledge he possessed, that the utmost danger awaited us if we should work into the harbor, as was at present our intention. Upon learning this, the officer in command on deck received immediate orders not to proceed any farther in endeavoring to work the ship into the chops of the harbor.

The Reverend Gentleman then went on to state, that a few months after the ship *Duff*, Captain Wilson, in which vessel he had arrived at this island, had left, another ship had touched there, for water and refreshments, from which an Italian renegado had deserted, and secreting himself until the vessel's departure, still remained on the island. This man was possessed of a very insinuating manner, and had moreover taken with him, at the time of his leaving the ship, a musket, a quantity of powder, and some balls, by means of which he very soon so far ingratiated himself into the favor of the leading or principal chief, as to become a prominent director in the affairs of the island. It was upon this man's proposition, that the war with the natives of La Domineaque, which had raged for some time with all that savageness and barbarity peculiar to their mode of warfare, had been commenced; he had also instigated them to fight against another tribe, adjoining whose land lay the estate and place of residence of the chief which had brought Mr. Crook on board. This was at a considerable distance to the westward of the harbor, and was the spot whence they had paddled off to the ship, and where, in company with his friend, the chief, Mr. Crook had been keeping watch, anxiously waiting for an opportunity to carry their plan of escape into effect. It was in consequence of Mr. Crook's disapproving of the wicked plans and enterprises of this fellow, and because, as feeling it to be his solemn duty to his God, and these his fellow-mortals, he had protested against his farther leading them on in furtherance of his abominable practices, that he had become bitterly opposed to Mr. Crook, and was the cause of all his painful distress; to such an extent did this Italian's hatred for Mr. Crook lead him, that at last the principal chieftain and several of the petty chiefs, were (by him) induced to watch for an opportunity to murder Mr. Crook. The natives were the more ready to submit to this Italian's management, because of his possessing the musket, powder, and shot; the wonderful superiority of this instrument in battle over their own arms, leading them to believe he was invincible; and with his aid, he persuaded them that they would not only be enabled to conquer all the tribes in both the islands, make them to be subjects, and pay tribute to their principal chief, but would furnish sufficient means for them to take and destroy every vessel that would hereafter stop at their harbor, and possess themselves of all the iron and valuables: but before any thing of this kind could be done, he was exceedingly solicitous that they should massacre Mr. Crook.

This gentleman, while alone among the Indians, had, by his kind behavior and regard for their well being, secured the affections of many of the chiefs, but

THE MISSIONARY SHIP "DUFF," CAPTAIN WILSON
From an engraving published in London in 1805

NUGGOHEEVA, MARQUESAS ISLANDS

From an engraving in Porter's *Voyage in the South Seas*, London, 1823

none were so warmly attached to him as his friend who had brought him off: this man being their first war chief, a station giving its possessor much influence and weight in their counsels, and second to none in the tribe, except the principal leader, had often boldly confronted them, and exposed his own life to save that of Mr. Crook, which was daily in imminent danger, from the ambushes and snares that were laid to entrap him, continually changed, and suited as they were to destroy him, as he was found to be more or less attended by his friend; unable, however, to succeed in their wicked attempts, at the same time well knowing that both Mr. Crook and his friend were acquainted witht the plan of operations, and therefore sure that if either of them should succeed in getting to the ship, their hopes of cutting her off would be at an end, by the disclosures they would make, word had been sent to them early in the day, by which both were informed that it was the desire of the principal chief, that neither should go on board the ship (which, according to a custom among the chiefs, amounted to a taboo), as he had concluded to go and see the captain himself. In order to secure the more faithful obedience to this mandate, and watch over them, a petty chief frequently called from the harbor (using as a cover for his main errand), to consult on the plan, report progress, and counsel and advise with the chief at his residence; here had these two, the moment our ship first appeared in sight, kept a lookout, and so soon as those two chiefs, who had remained so long on board, acting partly as pilots (one of them whom we now learned was the principal, the other one of his counsellors), were known to have left the ship with the other natives; they embraced the opportunity offered by the thick rain squall, and put off; their risk was great, for death it was thought, would have been the certain lot of both, had they been intercepted. There really appeared to be a particular Providence attending us, and I am free to acknowledge, that afterwards I felt self-condemned, for having suffered my mind to be chafed by the obstruction experienced in our advance, from those squalls and gales of wind, and which had been, by preventing our getting into the harbor, the means of our preservation; as most likely, had we this day so anchored, all would have been cut off and massacred. Our ship was to have been their first victim, and from her small size, would have been the very one to be desired, as they were much more likely to succeed upon her than against a larger. We now saw distinctly, the reason why the two chiefs were so earnest in their solicitations to have us enter their harbor; as also why the productions of the island were so very scarce: the renegado had, in fact, been completely successful in engaging the head chief to take the very prominent part in his plan of operations, which he was then acting out, and as there was not on the island, at this moment, a supply for our ship, the promise that an abundance should be given us, was but a portion of his share in the villanous scheme.

The mode by which they expected to succeed in the capture of our vessel, as we now learned, was, when night shut in, to send off swimmers and divers with the end of a rope, to be made fast to the ship's rudder hangings, still keeping the other end on shore; the ship's cable was next to be cut off under water, and so soon as this was accomplished, the natives on shore were prepared to haul away on the line, and drag her with what force they could muster, on shore. The Italian, in all this design, had proved himself a cold-blooded monster, and a man altogether void of any humane feeling. He had made the people believe that their success was certain, and lest a vestige should have remained, by which the affair could ever have been discovered, all hands were at once to have been destroyed, and the vessel burned.

Thus they hoped to gain two points at once; suffering nothing of the vessel, or any of the crew, to remain in existence, and secure all the iron, an article which they held in great estimation; the powder, cannon, and fire-arms, were to be kept for the purpose of more easily taking the next vessel that should arrive, as well as a means by which they were at once to be made more powerful than the other tribes, whom they could then safely go to war against; thus becoming the most wealthy, most powerful, and of course the greatest of all the islanders in the Pacific. The whole plan seemed so them as easy to be accomplished; great dependence was also placed upon the divers and swimmers, and very few of the natives but were proficients in this business. The utmost silence too was to be preserved, and thus they expected to avoid detection from the watch on deck: when the cable was cut, by hauling the ship astern, until she went aground, they were made to believe she would immediately keel over, so that her great guns would become unserviceable, and their superiority in numbers make the capture an easy one.

Thus engaged in conversation, the time had passed very rapidly, and it was near sunset when the chief, getting more and more uneasy, was informed by Mr. Crook, that as he had now found an opportunity to return to his country, it was his duty to embrace it, and therefore he could no more go back with him to the shore. The chieftain was much distressed at this announcement, and expressed himself as fearing that he should not long survive the separation; but knowing the danger Mr. Crook was exposed to in going again into the canoe, he could not ask him so to do; yet hoped it would not be many moons before he would again see him back to their country, together with Captain Wilson, by which time their wars would be ended, and all things put right. Mr. Crook was much affected by the chief's attachment for him, and replied, that if Heaven was so pleased, he hoped ere long to have the happiness to take him by the hand again, reminding him at the same time of the promise* he had given to notify the captain of the next vessel, and of all others that should come to their island of the danger, in time to prevent their receiving the smallest harm to crew or vessel; this he promised most faithfully to perform.

When we had descended at first into the cabin, I had taken from my pocket my pistols, and laid them on the cabin locker: the chief, after minutely examining their locks, handles, and barrels, inquired of Mr.

*True to the promise he had given, the captain of the ship *Buttersworth*, the next vessel that touched at this island, received sufficient early information to avoid the danger, as the author has been subsequently informed.

Crook what they were good for, particularly into the manner in which they were used; being satisfied in all these particulars, he next was very solicitous to have one of them presented to him. Though preparing to make him a suitable present, I was not willing to part with either of these, and therefore answered, that as all the arms, powder, and guns, on board the ship, were *tabooed** by the superior chief in my country, none of them therefore could be taken out of her to be given to any person. The explanation was highly satisfactory, and the respect they have for any thing tabooed, appeared to close all his interest in anything about the pistols.

I now proceeded to lay out for him such articles as were deemed the most acceptable, such as a few axes, some hatchets, knives, razors, with an assortment of small cutlery; adding along with them, to complete the variety, a parcel of beads and glasses, until Mr. Crook, who, when requested to state what would be most highly prized as a gift, had declined so to do, leaving it altogether to myself, now said there was a great abundance, and a present so ample he did not hesitate to say would produce for the chief a good reception upon his return to the shore, notwithstanding his instrumentality in securing the escape of Mr. Crook. When told that all these things were to be put into his canoe as a gift for himself in consideration of his being Mr. Crook's friend, and of the promise he had given to do his best to prevent any vessels which should hereafter touch at their island receiving any damage, he was greatly elated, boasting that now he

*Prohibited, or forbidden.

was the most wealthy among all their people, more so than even the head chief, whom he could now boldly tell, that the principal cause of all the distress among them was in consequence of their keeping such a bad man there as the Italian deserter was; and moreover, that so long as he stayed, they could never expect peace, plenty, and comfort, but would surely have contentions, and bloody wars, and strifes.

The parting scene between these two friends was truly affecting; and such a one as might well be supposed to have taken place, where, on one side was a heart duly capable of appreciating the attentions so long received, and the imminent risk which had been run in his behalf while effecting his escape; and on the other, was a child of nature, possessing virtues and feelings that would have been creditable to a civilized being, and who felt that he was now parting with the only man who had ever given him so much knowledge about his eternal welfare, who had ever taught him to see the little claim matters connected with this life had to his attention, compared with those beyond the grave.

When yet some fifty yards distant from the ship, paddling for the shore, the chief stopped his canoe, and called out to Mr. Crook to return in a few moons at farthest, or he should not live to see him again; then giving the friendly flourish, with his paddle, he continued his way for home.

From Mr. Crook I received information of the existence of another group of islands, four in number, which he mentioned had been recently discovered by one of our countrymen, in a ship from Boston, and called the Washington Islands, laying in a northwest direction from these. He was very earnest in recommending our immediately sailing; having understood that their language was very similar to that spoken by the natives of the Marquesas Islands, which he was well acquainted with, and well qualified to speak, there could be no doubt but Mr. Crook could also understand and speak that of those new islands, at all events sufficiently well to act as our interpreter, in procuring water and refreshments, as we should want them; which subsequently was found to be the fact.

There being no prospect that we should obtain any supplies at this place the ship's course was accordingly directed for the Washington Islands. On returning to the cabin, after seeing our friend the chief off, I again requested Mr. Crook to accept a suit of clothes from the number handed out, for he had, ever since coming on board, been still attired in the native costume; with this wish, after many acknowledgments, he was pleased to comply.

It had been observed, that at the time when the natives were very numerous around the ship, then laying off Resolution Bay, some of them would take fish, from four to six inches in length, just as they were caught, and eat them, beginning by first biting off the head, so on by a mouthful at a time, until the whole was eaten, or they had finished. On mentioning this to Mr. Crook, at the same time asking whether it was not customary for them to cook their fish, he replied, if the fish was large, and their provisions were plenty, they did cook, but owing to their wars, and the attendant famine, their sufferings for provisions, which were now very scarce, had been great; concluding this to be the case with those we had seen; adding, that himself had been driven to so great distress at times for food, as to do the same thing; this he was obliged to do at the first, so soon as he had caught the fish, or it would have been taken from him; and added, that while eating one of these small raw fish, he thought he had never tasted a sweeter meal: he said it was a fact also, that the natives, when pushed by famine, would make use of all the art they possessed, to get one of their enemies into their hands, for the purposes of food, it being altogether out of his power to put a stop to so inhuman and horrid a custom.

May 23d. Ever since leaving the Marquesas Islands, we have had heavy rain squalls from the south-eastward, with calms at intervals. At 1 P. M. had sight of the southernmost island of the Washington group, bearing W. five leagues distant. At 3 P. M. the eastermost island was seen in the north-east quarter, and half an hour after, the northermost and largest island of the group, was in sight, bearing N. W. half W. distant about eight leagues. Steered for the eastern end of the southernmost island, and at 9 A. M. while moving along its northern coast, under a favorable breeze, we opened a bay, whence several canoes, some of them very large, others again small, came off to us. These large double war canoes were similar to those at La Christiana, especially the one which had brought off the head chief; like it, the bows of these were ornamented with a number of human skull bones, which Mr. Crook stated to have belonged to their enemies, whom the chief, the owner of the canoe, and his warriors, had slain in battle. I had the satisfaction to find that Mr. Crook could carry on a conversation with this people, as fluently as with the Marquesas islanders, their language being nearly alike: they appeared to be greatly surprised, as well as pleased, to hear Mr. Crook speaking in their language, and were very anxious to find out where he had "catched it,"— to make use of their own expression — wanting to know where he had come from, thus to talk as one of themselves. On his inquiring whether or no there was a harbor up the bay, the chiefs answered in the affirmative; but none of them or the other natives could be induced to come on board, notwithstanding there was one there who spoke their language. Many of their canoes were armed, for such I supposed they were, from the heaps of round stones, war clubs, and spears, that were in them.

We now stood with the ship into the bay, but had scarcely got within the chops of its headland, by which it is formed, than we were becalmed, and a heavy rolling swell hove the ship up the bay; meanwhile, we were endeavoring to keep up a trade with the natives in their canoes, giving in exchange for their products, toys, glasses, beeds, buttons, &c.; iron they would not accept, holding it in the most sovereign contempt.

It did not appear that these people had ever seen, or been visited, by a civilized being before, nor did they show that desire to obtain iron, which all the other islanders so strongly manifested, not being willing to receive it even at the rate of a hatchet for a single cocoanut, which a small bright button would readily purchase; indeed, the bits of broken bowls, pitchers, or crockery of any kind, which the steward still kept on board, appeared to be far more valuable in their estimation, than any other article in our possession.

By the help of the glass, the surf was seen to break high on the coral reef, which bounds the shore of this bay, and on my observing to Mr. Crook, that the ship had got hove so far up into it, as to render it highly necessary to ascertain from the chiefs whereabouts their harbor was, and what was the depth of water they there had; in answer to his inquiry, they replied, it was up at the head of the bay, and where they hauled up their canoes: as it respected the depth of water, they stated that the bay had no bottom, without the reef, but within the coral reef, there was a good beach, upon which they were in the habit of hauling their canoes. To discover this harbor had now become an important point with us, and not at all to be trifled with; no time was to be lost. We therefore manned one of the boats, for the purpose of towing the ship's head around, in order to head the rolling sea; and when this was accomplished, kept her still towing out of the bay, at the same time sending another armed boat to sound out the head of the bay. In this duty she was some time engaged, without the officer in command giving any signal, by which we could understand he had found an anchoring spot, nor could the first boat do any thing more than keep the ship's head sea-ward, for as to moving her forward, it was soon found to be impossible.

The natives were all this time entering the bay, and a vast number in their canoes were continually coming round its head, from other parts of the island, some of which were gathering round the sounding boat, so that in a little while the bay was quite filled with them, especially in the immediate vicinity of the ship.

I now observed by the countenance of Mr. Crook, who was constantly near me, that there was something which tended to make him uneasy: nor was I long kept waiting an explanation; for whispering, he gave me to understand, that from the conversation passing among the natives in the several canoes, and from their exclamations, he judged they were not peacefully inclined, and as many of the canoes had gathered together between the ship and the sounding-boat, for the purpose, apparently, of cutting her off on her return, it was thought to be most prudent to raise the ship's ports, and run out her guns, which had been previously charged with a single shot. A signal was also made for the second boat instantly to return, and for the purpose of opening a passage for her to the ship, a musket was discharged over the heads of the natives in a direction to let them hear the whizzing of the ball, see it strike, and then make the water fly up; at the same time running the carriage guns out. The effect was as expected; for they instantly drew back, and gave the desired free passage for our boat, which, much to our satisfaction, now pulled safely alongside. The natives now sounded with their conks, the war notice, and accompanied this with a succession of the most deafening shouts. Mr. Crook called to some of the nearest chiefs, and counselled them to keep at a distance from the ship, nor suffer any one to throw a single spear, reminding them of the effects they had seen the little gun produce upon the canoes which were in the way of the boat, and inquiring what they could expect, when the fire and thunder was let out of the big ones, but to be all destroyed, with their island, adding, as he pointed to the guns just run out, the captain, as they might see, was determined to destroy them and their island, as far as the thunder and fire from the ship could do it. After expressing their astonishment at such wonderful power, the ship, they said, must certainly have come from the clouds, and very soon after paddled off to a more respectful distance, but did not, however, cease their shouting, or blowing their war conks. When some of the nearest chiefs beheld the bright blade of a broad-sword glittering in the sun's rays, they declared, one to another, that that battle axe must have come from the sun, it was so dazzling.

The officer who had been out sounding, reported that with fifty fathoms of line, within a cable's length of the reef, he had not been able to get bottom. Finding no progress was made with one boat's towing, the other was put at the same duty, and our sweeps were got out to assist them. At times, a little would be slowly gained against the rolling swell and in draught; then again, in a few minutes, there would come a trio of mammoth rollers, so that in spite of all our exertions, her headway was sometimes stopped, and sternway given to her. The islanders gathered in groups on the shore around the bay, on the rocks of the coral reef, and many hundreds of them still in their canoes, were intently watching all these contrary movements, and whenever the ship fell back were sure to raise a tremendous shout, making the bay ring again with their uproar.

After several hours spent in this laborious yet necessary exertion, and when, by every effort that could be made, we had been able to gain but a little over a mile in our course, and when the strength of our cheerful and spirited crew was nearly exhausted, we were blessed by a faint flaw of wind from the eastern head of the bay; not a moment was lost in trimming every sail to make the most of it. In a little while she began to move at an increased rate, and finally effected an offing in safety. This we called Escape Bay, in consideration of our very narrow escape therefrom.

Kind Providence had again interposed in our behalf, in thus preserving our vessel and lives from so painful and dangerous a situation; and to the Rev. Mr. Crook, who was ever attentive in giving early notice of any newly discovered evil intent, or movement indicating hostility on the part of the natives, and who was ever at my side, I am much indebted. Throughout the whole of this trying season, the good order and discipline of our crew were clearly manifest: there was no murmuring of discontent, no want of confidence in their leaders, and except the giving and passing of necessary orders, all were silent. To the above, together with the promptitude with which every duty was executed, in my opinion is our extrication attributable. Our ship was small, and managed comparatively easy; but on the contrary, had she been a larger and heavier vessel, no effort or invention of man could have saved her, and the lives of all on board, from that destruction which so fearfully threatened us. Owing to the inset, and heavy rolling sea or swell, she would have been undoubtedly hove on the coral reef, and dashed to pieces by the tremendous breakers, in less than half the time employed by ourselves in getting clear; the massacre of such of the crew as should escape to land, terminating the catastrophe.

The situation of that celebrated and much lamented voyager, La Perouse, in his two large frigates,* must, in all probability, have been similar to that of ours, and no doubt was the cause of his loss. The flaw of wind that assisted in our escape, did not, at the time, ruffle the surface of the water over the distance of fifty yards astern. It was very gratifying to observe that even among the seamen, there was an apparent consciousness that their preservation was altogether owing to the goodness of that being who had deigned to assist them; and as they hung around the Rev. Mr. Crook, every honest heart seemed to wish him to declare their heartfelt thanks to the Almighty for it.

The former impression, that no ship had ever visited this island before ours, received additional strength from the fact that we did not observe or meet with any iron or beads, &c., among the natives, neither could Mr. Crook learn from them, in all their many conversations, that they had seen civilized beings before: the ship formerly spoken of, must therefore have visited another island of the group, and not this. He had, however, procured from some of the chiefs the

*The first vessel sent to the Pacific under the agency of the author, in 1803, after a cargo of sandal wood, was of a full build, and met this lamentable fate. All on board either perished by drowning, or as they gained a foothold upon the rocks of the coral reef, were massacred by the natives. This was ascertained to be the fact from information subsequently obtained, as will be seen by a reference to the voyage of the brig *Union*.

name of their island, and had written the same down in the ship's log book: this according to their pronunciation is, *Hooapoah* or *Wep'oo;* the name of the easternmost island is, *Hoo-a-ho'o-na;* and the northernmost one, then in sight from the ship, and the largest of the group, thus, *Nug-go-hee-va;* the westernmost, and smallest, they called, *Fet-too-e'e-va.*

Our stock of water had by this time got reduced to a very limited quantity, and imperiously required that it should be recruited; but to think of obtaining an additional supply of so important an article at this island, in the present hostile spirit of its natives, was entirely out of the question.

Matters and things being thus situated, it was thought most advisable to pass the night in standing over for the south coast of the large island Nuggoheeva; the ship's sails being accordingly trimmed to accomplish this object, we stood to the north upon a wind, and at daybreak were about two leagues from the south shore of Nuggoheeva, but well towards its west end; therefore stood in, and commenced working up along shore in search of some harbor. During this time we were favored with a fine beating breeze, and smooth sea.

CHAPTER XI

AT NUGGOHEEVA ISLAND

MAY 25th, 1798. We were employed in working the ship along shore to the eastward. When standing upon our in-shore tacks, and near to the same, the natives would put off in one or more canoes, and come within speaking distance of us; they appeared to be, though rather fearful, a very sociable sort of folk, asking as many questions as they saw fit, and giving answers to our questions as well as they were able: to the Rev. Mr. Crook's inquiry, if there was any harbor near to where we were, their unhesitating reply, invariably, was yes, a very good one, where they hauled up their canoes. We now discovered that they considered every place where their canoes could be hauled on shore as a harbor; this must likewise have been the meaning of the chiefs at Wepoo, and when they said there was a good harbor, they intended to tell us there was a good beach, where boats or canoes could be readily drawn up; and our not sufficiently understanding each other, was the cause of our getting into that difficulty, and not any intentional misleading on their part: every place where they could haul up a canoe, they thought was what we called a harbor.

Notwithstanding we had displayed a white flag as the signal of peace, and as such understood by the natives, and all Mr. Crook's endeavors to prevail upon them to venture on board, yet it was to no purpose; that they comprehended what he said was plain, from the lengthy conversations held between them; although they were carried on at a speaking distance, still not one would come even so near as alongside; the guns we had been careful to run in, as well as to close the ports. From this want of confidence, the little traffic we had was necessarily carried on in the same distant and inconvenient manner; whatever we had to dispose of, being made fast to the end of a line, and thrown over the stern to them, they managed to get, and very honestly would make fast in payment, what they conceived the same to be worth; thus beginning and concluding our exceedingly small business. A course of conduct so reserved, we were much puzzled satisfactorily to explain: it was not solely restricted to this body of the natives, for the same temper was observed among others.

At 11 A. M. when abreast of a bay, where to all appearance we had at last found a harbor, three canoes came off to the ship, laden with cocoa-nuts, bread fruits, and the like articles; yet nothing could induce the natives to come on board, though our exertions were redoubled, and every plausible contrivance was resorted to for the purpose. They gave us to understand that ours was not the first vessel they had seen; which was confirmed by some of their company showing a small number of beads, very ingeniously fastened to the tusk of a hog, and handsomely polished. As we were on the point of hoisting out a boat, in order to examine more particularly, whether or no a good and commodious harbor was not somewhere close by, the man on the lookout at the mast-head, called out, that he had just at that moment, seen what had the appearance of being a larger bay, a little farther to the eastward: going aloft, and taking a look at this with the glass, the prospect promised a much more favorable result to our labors, than the one we were now on the point of examining.

We accordingly without much ceremony of leave taking, left the canoes, and made a board with the ship off, in order to fetch in near to the other bay, which was successfully accomplished. An hour after meridian, as the ship was hove to abreast of the mouth of this harbor, a large canoe paddled off to meet us, from the shore in which was an aged chief, whose white locks gave him a very venerable and interesting appearance; there was also in this canoe, some thirty natives, each one being a paddler. The old man quickly displayed a white flag, together with a green branch: these tokens of friendship and amity we answered in like manner, by holding up a white flag, at the same time desisting for a while from hoisting out the boat, which we were about doing, expecting that all the information we desired, could be obtained from this old chief; but no, he was proof against all we had to say, most likely to bring him alongside for this purpose.

Around the ship he several times paddled, keeping at fifteen or twenty yards distance, and taking all due caution not to come too near; this through with, the old chief next proceeded with giving a specimen of native oratory, the purport of which was to present a cordial invitation from the king for us to come on shore. After waiting with the utmost patience for the termination of this haranguing and manœuvring, and completely tired out with staying for him to finish, however uncourteous it might appear, I ordered the boat to be hoisted out, and mentioned to the Rev. Mr. Crook, that we would now see, if by stratagem we could not get this aged chief on board, provided, however, that it could be executed without any accident, or unpleasant occurrence; inquiring, at the same time, whether he was willing to assist in accomplishing the same, by taking a seat in the boat with the officer. To this he most readily consented; when the officer received these directions: in the first place, to sound out the bay for suitable anchoring ground, and if the canoe should follow him, not to appear to notice her, but so soon as he had finished sounding, to embrace the very first opportunity that should present itself, and endeavor to lay his boat, which was a remarkably quick rowing one, at once alongside of the canoe, and seizing upon the chief, bring him instantly on board, if this could be completed without harming him. The boat was quickly prepared, and manned by a select crew, the Rev. Mr. Crook taking his seat in her alongside the officer: meanwhile a conversation was kept up between Mr. Crook and the old chief, during which the boat put off, steering directly for the bay: as was expected, so it turned out, for so soon as she had gone, the canoe followed, keeping close by her. It was necessary for the boat, in the performance of the duty assigned, to cross and recross the bay, going backwards and forwards very frequently, so that she kept gradually gaining towards the shore; this, when the natives perceived it, increased their confidence to such a degree, as finally to bring their canoe so near as to interfere with the boat's oars, Mr. Crook all the while keeping up a friendly conversation with the chief: matters proceeded in this familiar way until the officer had finished his soundings, and made the signal of having found a suitable anchorage; the canoe was at this moment within ten or twelve yards distance, which the officer observing, gave the word to the crew, who were all ready, and in a moment the boat shot alongside the canoe; this manœuvre was executed so promptly, as completely to surprise the natives; not so much, however, as to prevent every man of them jumping overboard, leaving their aged chief, who sat trembling with fear, to get out of, or along with, this unexpected difficulty as well as he could.

Mr. Crook soon informed him, that the captain, the head chief on board the ship, had sent him an invitation to come there, and he would soon be convinced that we were their friends; adding also, a request that he would not be alarmed, as there was no harm intended to him, nor should he suffer the smallest injury. With this assurance, the old chief replied, he would go with the boat, but first handed into her a green branch and a small pig; these, said he, are the emblems of peace, and if they are accepted by the head chief on board the vessel, then I shall think you are my friends. Mr. Crook informed him, that we understood the custom, and that they would be accepted; at the same time getting into the boat, followed by the chief, who, still trembling with fears and doubts, was there seated between Mr. Crook and the officer; these fears were in some degree diverted, when in compliance with Mr. Crook's request, he looked up and saw the ship under easy sail steering towards them, for so soon as the signal of her having found a suitable spot for anchorage was perceived by those on board ship, we immediately bore up and stood in towards her. As we met, the ship was brought to, in order to receive the chief on board, Mr. Crook requesting me to receive the pig and green branch, the better to calm the chief's uneasy state of mind, and pave the way the more securely to bind his affections hereafter; accordingly on receiving him at the gangway, the old chief presented in the first place the green branch, accompanying this act with a short address; after which, doing in like manner with the pig. When on deck, he insisted upon paying homage, but such I informed him, while raising him from this posture, and handing him to a seat on the quarter-deck, was not the manner of saluation when friends meet friends in my country, and as such I hoped we had now come together, adding, that I myself was but a chief like himself: yet, said he, as I was given to understand through the interpretation of Mr. Crook, there is this difference, you came from the thunder in the clouds, and are therefore more powerful than even my king.

The natives had immediately returned to their canoe, when they found our party had left her floating about, and paddled on after their chief, taking the utmost care, however, not to get too near the boat. After sending the boat out at a proper distance ahead, to sound the way, we filled away on the ship, and steered in after her, passing in between the two high, but small, round islands, which, because they so nearly resembled each other, we now named the Sisters, and entering a spacious bay, hauled up towards a beach on its eastern side, and anchored in twelve fathoms of water, in a fine harbor with clear ground.

The old chief, Mr. Crook, and myself, had returned to the deck after going through all parts of the ship where it was possible conveniently to get, showing and explaining all the uses and purposes of the different warlike apparatus, the cabin, with its furniture, and in short every thing, the more minutely, because we were determined to leave no effort untried that could possibly secure his confidence; yet there we were, unsuccessful after all. The Rev. Mr. Crook, who had noticed that the chief seemed to have a great fancy for the robbin I then had on, mentioned this to me: it was a short roundabout, made of red flannel, with tape strings of the same color. This I instantly took off, and gave to the chief, who very speedily made himself perfectly at home, after putting it on; he was at once vastly elated, and appeared quite another man. It was truly laughable to see the old man strutting first up and down the deck, with this great acquisition on; then marching aft, the better to enable his attendants in the canoe (who had ever since followed on, keeping at about four or five rods distance astern) to admire his person, thus ornamented. They laughed heartily, and shouted for joy, at this prodigious fine display of their chief, and were now of a mind to be wondrously sociable. All but two of them came at once on board, after the chief had called to them to paddle up and make fast to the ship, while he went on, as well as the frequent admiration of his prize would allow, to state who he was, what in the goodness of his heart he would do, &c.; the sum of all being, that he was the young king's grand-father, and now, we should have plenty of hogs, bread-fruit, yams, &c. While this was going on, the canoe already spoken of was the only one to be seen, at, in, or about the bay; but at this moment, two large canoes were seen coming from towards the western part of the bay, and as they came alongside were found to have brought with them four fat hogs, some bread-fruit, cocoa-nuts, yams, bananas, sugar-cane, &c., as a present from the young king. In return for this, there was sent to him two axes, some hatchets, chisels, looking-glasses, bright buttons, and beads; and to our new friend, the old chief, who took it in charge, we made a present of a hatchet, knife, some pieces of iron hoop, some fish hooks, nails, and beads; not forgetting the two chieftains who had brought the king's gift, giving each something suitable.

We were at this time much in want of fresh water; it became therefore, highly important for us to be looking around to see whether any could be obtained at this place. On Mr. Crook's asking the aged chief where we could find a supply, he replied that the "biggest" water was near the king's village, off towards the western part of the bay, and if we were so disposed, he was perfectly willing to go along, and show whereabouts it was. This generous offer we of course readily accepted: after having previously furled the ship's sails, and putting into the boat two small iron-bound casks, together with some small matters for trading purposes, the party then started off, the Rev. Mr.

Crook willingly accompanying them with the old chief in his canoe keeping alongside the boat. On their return with the casks filled with fresh water, they reported the place where it was obtained to be a beautiful small stream; though when the casks were filled and bunged, they were obliged to swim them through the surf to the boat, which lay at a grapnel, it being thought unsafe, in consequence of the roughness and rocky shore, to attempt a landing with her. The natives they found to be very friendly; and many, anxious to do something, readily volunteered to swim the casks off to the boat. A great many were constantly bathing in the river: this was the case around the ship; some swimming a little, then seizing hold of some part of the vessel to rest for a fresh start, looking, as they hung around her sides, much like a flock of blackbirds upon a tree; others in canoes, some near by, but more farther off, were quite content with this outward view of so strange a craft, keeping their tongues going at a merry rate all the while, each one anxious to tell all he know, and more too. The current of intimacy between us, was now as difficult to be stopped, as it was formerly to be acquired; and in our present situation, it was thought most prudent to permit only such of the natives to come on board, as were known to be the attendants upon the several chiefs.

With a view to enter into an arrangement by which our ship's water might be filled, the Rev. Mr. Crook proceeded in the ship's boat to the shore, in order to find out our friend, the aged chief, and invite him to return with him; for from this friend we hoped to obtain a taboo to be put on in the morning. The invitation was not only accepted, but the old chief (whose name for the first time we now learned to be Tearoroo) came accompanied by Toohoorebooa, the regent chief, and uncle to the king, bringing along with them a second present, consisting of hogs, yams, bread-fruit, bananas, sugar-cane, tarroos, &c. which the young king Paeroroo had sent. They refused, however, to receive aught in return, giving as a reason, that the king did not wish it, for he had plenty of such articles as he had already sent, to spare. Toohoorebooa now requested to exchange names with me; and Tearoroo joined in the wish, the regent concluding, when we had thus exchanged, with observing, "now we are true friends," this introductory ceremony being got through with, Mr. Crook went on to state what our wants and desires were: that having many casks to be filled with water, we could not do it, because his people were so much in our way as not to allow us to make any despatch in filling, or to get them off when filled. Here he interrupted our complaints, with the request that we would knock them in the head if they did not keep out of the way in future. This summary mode of procedure, Mr. Crook informed him, could not be consented to, and went on to request as a particular favor, that he would consent to put the river and bay, during the following day, under a taboo; stating as a reason for this, that the men from the ship did not like to fill their casks when the natives in great numbers were bathing above them in the river. This the regent consented to do at sunrise, as well as to send messengers up the valley and have the river tabooed up to its source; but added his desire to remove this when the sun went down to the tops of the western trees, so that the people might bathe before dark. This provision it would have been very ungenerous for us not to have agreed to: we therefore, of course, immediately assented to it, stipulating also to notify him at the time when we should have finished.

This taboo or restriction went even so far as to prohibit all natives from swimming in and about the bay, forbidding the use of their canoes, unless our permission was first obtained, excepting always the king's. The regent was, in addition to this, to furnish a sufficient number of good swimmers to take the water casks to the shore, and after they were full, to retake them through the surf to the boat.

After completing this arrangement, the most perfect understanding subsisting between us, these friendly chiefs, at just about sunset took their departure; and all the natives, who had many of them been a great deal of the time much in our way, in compliance with these chiefs' orders, followed suit; a good riddance for us. We then proceeded to make every preparation for the more certain getting through our watering business in one day if possible.

On the passage, and at the same time when we had layed our set of running rigging, we had also layed a quantity of nine-thread rattling rope, for a boarding netting: this, when fitted to its place, going all around the ship, and triced up to the tops, jib stay, &c. was twelve feet in height above our railing, and was a means of safety against surprise during the darkness of the nights. Yet as a still farther measure of security, one sentry was placed at the heel of the bowsprit, one on each side, abreast of the fore and main masts, and one on the taffrail, who called out, commencing with this last, every thirty minutes, "All's well!" an officer on the quarter-deck having charge of all. Thus I considered we had taken every possible precaution to guard our little ship against surprise. This order of things, our guns charged with a single shot and a bag of musket balls, all ready and run out, was constantly kept up during our stay here.

The following morning at sunrise, having prepared everything the night previous, as was before stated, sent the boat to the shore with a raft of empty casks. The Rev. Mr. Crook volunteering to attend on shore, where, from the friendly disposition evinced by the chiefs, he was fully satisfied not the least danger was to be apprehended; the cooper and two of the hands were sent to fill and bung the casks. On their coming to the mouth of the river, the swimmers which the chiefs were to furnish, were found, according to agreement, all ready to go to work; they had been some little time waiting for the boat, and now that she had come, the watering was immediately commenced. The taboo, it was also to be perceived, was in full operation, as not a native was seen in the river.

Several of these swimmers, in their willingness to render assistance, and to please their new friends, after obtaining permission from those in authority, did swim a cask off to the ship, a distance certainly not less than three quarters of a mile, receive a board nail as ample compensation, and then swim back to the shore for another cask, with as much spirit and earnestness as if they had a more important prize depending upon their exertions. So unremitting were the efforts of our friends, that by 5 P. M. we were abundantly supplied with fresh water, a quantity sufficient to last during our passage to Canton.

Having thus secured a good stock of one of the chiefest articles necessary for our long voyage, the regent chief Toohoorebooa was accordingly so notified, and the taboo forthwith removed. In a very little while after this restoration to one of their rights and privileges, of which for the time being they had been on our account deprived, multitudes of the natives surrounded our ship; some in canoes, others swimming or floating alongside like a shoal of porpoises, bringing along with them figs, fowls, bread-fruit, cocoa-nuts, yams, tarroos, sugar-cane, and the like. So many customers, of course, made as much business for us just then, as we could conveniently attend to; bits and scraps of old iron were in great demand, and had now as much value attached to them as ever they had: these, with porter or wine bottles, were taken unhesitatingly in payment, and were held in much higher estimation than aught else we possessed. Sometimes the natives were gathered in such numbers around the vessel, hanging to her sides wherever they could hold on, as to give our little craft quite a rank heel, waiting or rather seemingly determined not to let us forget they were there for their turns, and incessantly jabbering at a rate sufficient to turn one's head.

Our friend, Toohoorebooa, the regent, accompanied by Tearoroo, our first and most aged acquaintance, came on board to make a visit, but more especially to inquire how we were satisfied with the manner in which the watering business was performed. On expressing my acknowledgments for their kindness, in thus securing to us the assistance by which we had been so greatly furthered therein, they were quite pleased, as well as to hear we were satisfied with the quantity and quality, and in conclusion offered, if we wished it, to put the taboo on again the next day. From these chiefs I received a pressing invitation to go on shore, and make their king a visit, in return for that which had been made to us. To this there could be no objection; it would also be a farther means of securing their confidence: still, I replied, it was necessary to consider whether I could or no, and in the morning they should receive an answer. Directly after sunset, the gun was fired for the setting of the night guard, whereupon the chiefs and natives took their leave.

The attention the Rev. Mr. Crook had given to the manners of this people, and a careful observance of their conversation, whether on board the ship, or when they were gathered around the watering party on the shore as mere lookers on, so far from inducing him to believe that any hostile plans were in preparation among them, had, on the contrary, strengthened his almost unlimited confidence in their pacific intentions; and he believed them to be a generous and open hearted tribe. Under this conviction, he was anxious that I should comply and gratify them, in reference to the above invitation, as he had no hesitation in saying there was not the least cause for fear; in fact, one would be perfectly safe, said he, to travel over the whole bounds of their tribe; every where they would be found to be the same in disposition and friendly feeling; but, continued he, they are acquainted with the usual custom of giving hostages, and will most cheerfully give up their chief men as such, rather than you should not come; and then mentioned, that he himself was greatly inclined to comply with their pressing solicitations to remain with them, as expressed to him while on shore during the day. To this portion of his remarks, I replied, that from the brotherly feelings and attachment subsisting between us, ever since he first came on board, not to make any mention of the great service he had rendered during this period, would make the separation rather hard, yet I was constrained to admit, that a correct perception of, or devotion to, the duty incumbent upon him as a missionary, in which cause it will be remembered he was engaged, ought now to govern him in his decision: however, the better to satisfy himself as to the friendliness of their intentions, and to ascertain what confidence could be placed in them, he concluded to pass this night amongst them on shore.

Early the next morning he returned on board, Toohoorebooa, the regent, being in company, his mind perfectly satisfied, as it respected the purity of their desires, and also that this was the field in which, for a short time at least, he had been appointed to labor, in a cause, for the complete and triumphant success of which all good citizens and honest men must raise their most ardent prayers.

From a custom held sacred amongst the natives, their young king was this moon tabooed from going upon the water, and with the queen mother, and other persons connected with the royal family, was very anxious to have a visit from the head chief of the ship at his royal house. A visit from such a person, Mr. Crook was of opinion would be of great service to him after our departure, by the influence a connection thus set forth would give him.

It was not a want of inclination to render all the assistance that lay in my power, to advance whatever might be suggested for the future good of the natives, that made me thus tardy in accepting such repeated invitations, but I did not conceive myself justifiable, or clear from censure, in detaining the ship, sacrificing my owners' time and interest, and that of my officers and crew, as well as my own, in gadding or roving about the country; it was not what I had been sent to do, and therefore was not right. I, however, replied, that could any possible advantage result from the visit, or could it be the means of insuring any benefit to those who might hereafter arrive at this island, I was willing to do whatever prudence might point out. Mr. Crook then stated his having overheard much of what the chiefs had said, while he was on shore, and from this it was evident that Tearoroo, the aged chief, and the brother of the regent, were to remain on board as hostages, until my return. With this arrangement, Mr. Crook thought it was best to agree, leaving directions with the commanding officer to keep them constantly in view, admitting but one at a time from the cabin, on deck; meanwhile, to the chiefs, he would state the necessity for their remaining in the cabin during our absence.

After this, I consented to go on shore, and mentioned 1 P. M. as the most suitable hour for me to make this visit to the king, directing the colors the while to be set on the ship. When Mr. Crook informed Toohoorebooa, the regent, of this arrangement, and that this was the cause of our setting the colors on the ship, he appeared to be very much pleased, exclaiming, *Vahvee! Vahvee!** and immediately arose, taking both my hands, and giving what is considered the most friendly mode of salutation amongst them, viz., a rather severe pressing of his nose against mine, concluding the whole with saying, that he was now happy, and would right away go on shore to make the king, the king's mother, and friends also, happy, by giving them this acceptance of their invitation.

Mr. Crook next suggested the propriety of preparing a medal of some kind to be suspended around the young king's neck; this he thought would produce a good effect. We, therefore, took a new bright metal plate, marking on it the ship's name, the name of the place and country she belonged to, viz., port of New York, United States of America, and finally where she came from; then making a couple of holes in the rim, through which we rove a yard of wide crimson ribbon, fastening the two ends together, the better to have it hang nicely round the neck, and rest on the breast; the whole we wrapped up carefully in a paper, so that no one should see it before it was presented to his youthful majesty. We had finished this as Tearoroo and the chieftain, brother to the regent, came on board, and stated the king had sent them to be hostages, until our return to the ship.

*Signifying, Welcome! Welcome!

At 1 P. M. accompanied by the Rev. W. P. Crook, the steward going along and bringing the medal, we put off from the ship, attended by a numerous company of the islanders in their canoes, and landed on a beach on the eastern side of the bay, where were assembled the regent and a number of other chiefs, together with any quantity of the natives; by this company we were received on landing. Two chieftains of the second rank gave their attendance upon the Rev. Mr. Crook and myself, and as soon as the order of procession was duly arranged, the whole moved forwards, preceded by Toohoorebooa, the regent, who wore on this ceremonious occasion, a most beautiful head-dress, made principally from the plumes of the tropic bird, intermixed with the feathers of other birds, the whole making a very splendid article; he had on also a breast-plate of the mother-of-pearl shell. Thus arrayed, he took the lead, which station he kept during the march. A row of eight chiefs, bearing long, black, and yellow rods, or canes, made of hard wood, having on one end bunches of human hair, marched on each side of us, while close behind us, came six chiefs in double files, and after all these, a vast multitude of the natives came slowly on in double Indian file, paying, however, no very great respect to their dressings. The march was around the north shore of the bay, in a direction for the young king's residence, and towards this we moved slowly, coming first to a small rivulet, and shortly after to the river where the ship's water had been filled the day previous. Just before arriving at the first, I found myself still moving, or rather moved, along in quite a comfortable manner, for without giving the slightest intimation of their intention, two of the natives had formed a sort of a seat, by clinching their hands together, then by striking the inner or hinder part of the knee, one is instantly made to set down on their hands; there is some risk of falling backwards, on receiving a blow of this kind, for rather than it should fail in making the knees bend, they give a pretty severe one. All this was so that no inconvenience should be felt by us in crossing these streams, or in getting our feet wet. I had received from the Rev. Mr. Crook a hint that possibly the natives would thus conduct themselves, and therefore was on my guard; not so the steward, poor fellow; he was most sadly frightened at being thus unceremoniously treated, and as a small return for the favor, gave them a most tremendous scream: from his station, which was a little behind us, he had had a fair view of the human hair which topped the rods of those chiefs at our side, and in all probability was thinking of the barbarous custom, which some of the South Sea islanders had, of murdering and eating their foes, when he was thus unwittingly seated, and perhaps he now conceived that his own hair was to decorate some other rod; but be this as it may, the effect of the scream was to make himself the laughing stock of all the natives, for all eyes had been instantly turned upon him, and raised as he was in the world, his countenance still made it very evident that he was much alarmed. This complete success of their manœuvre, pleased the natives prodigiously, and was the source of a great deal of laughter at the poor steward's expense, till some one good naturedly told him not to be

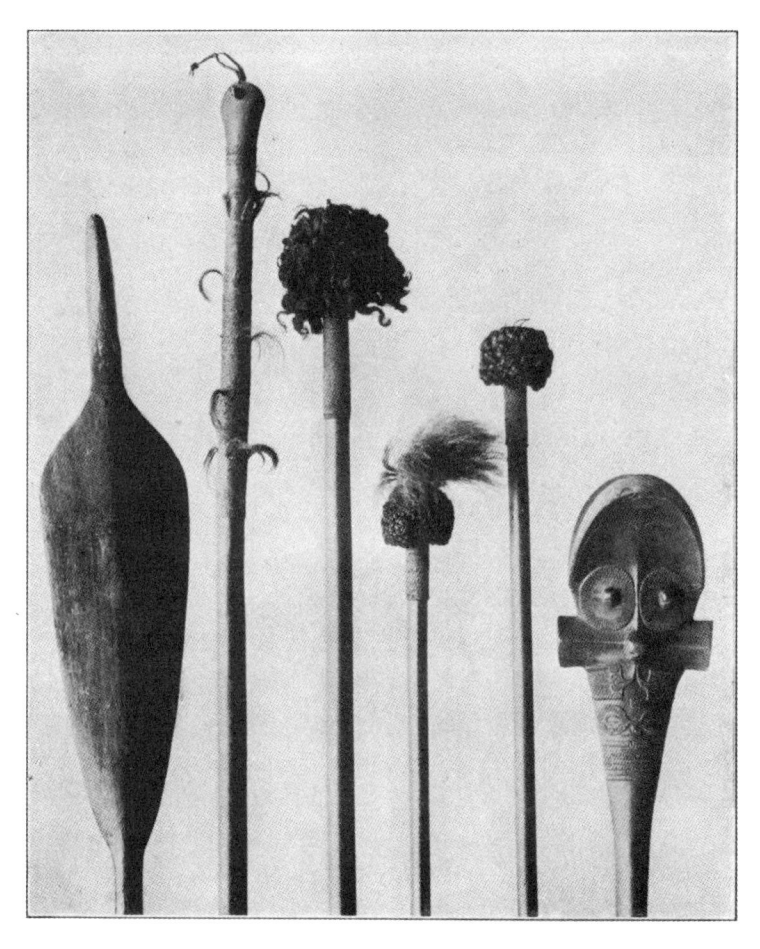

PADDLES, CLUBS, AND CHIEF'S STAFFS, MARQUESAS ISLANDS
From a photograph of the originals at the Peabody Museum, Salem

a. Blade of Paddle brought to Salem before 1802. *b*. Handle of long Paddle Club, before 1840. *c*. Head of Chief's Staff, human hair at end. *d*. Head of Chief's Staff, human hair at end. *e*. Head of Chief's Staff, human hair at end. *f*. Head of War Club, before 1817.

MODEL OF A WAR CANOE, MARQUESAS ISLANDS

From a photograph of a model brought to Salem before 1817 and now in the Peabody Museum, Salem

frightened, but to join in the laugh, if he wished them to cease.

After getting on the other side of this stream, we were set down very softly, to proceed as other folks do; the same piece of service being again rendered to us, when we came to the larger stream: the young king's abode was still some distance farther to the westward, and we accordingly continued our march. On the route, passing several groups of the islanders, who had been gathered thus together to see a sight so very unusual; these invariably fell flat on their faces, and remained in this position, until so much of the procession as were before the chiefs bearing the rods, including these also, had passed; they then rose, and came along or not, as they saw fit. We also noticed several groves of the valuable bread fruit and cocoanut trees. Around the bodies of some of the first mentioned, there were bundles of the coarse grass carefully wound, and on inquiring, what the object was in thus making a distinction between them, I was informed, that whenever the tree is so marked, all who belong to the district of a chief wherein there are any such, have a right to use as much as they please of the fruit; but, on the contrary, when the trees are without this distinguishing mark, they are in effect tabooed, or not to be molested: not a native could, upon any consideration, be induced to go to one of these and pluck its fruit. So great is their reverence for these simple laws, that they obey them with the strictest exactness; even a hatchet could not tempt any to get me fruit from an unmarked tree.

The king's dwelling, was situated in the centre of a

grove of bread fruit trees, having immediately in front of it, cocoa and palm trees, very handsomely arranged in rows, while on the outskirts, and towards the sea, there is an acre or more of very handsome grass, forming a beautiful foreground for the whole, and completing a view truly beautiful. The house was some eighty feet in length, by twenty in width: the walls were formed by placing rows of posts at equal distances from each other, both in front and rear, having a roof of the palm leaf, thrown over a stout sized ridge pole, and rafters of which appeared to be the bamboo, the whole being divided into four grand apartments of about twenty feet square, by means of thick mats, running from one long wall to the other, thus dividing the building equally. The two rooms at each end, are again equally divided, making six apartments in the mansion, the four on the end for lodging rooms. The floor of the entire building was of faced stone, and carpeted with mats of a great deal of finer texture than those hung for the dividing the body of the building into the different apartments. Extending the whole length of the front of the building, there were four rows of seats of faced stone, which answered the purpose of steps, while at the same time they were of so great breadth, that when seated on the uppermost, you could place your feet upon the next, without any inconvenience to those who are sitting there, and when well filled, look exactly as do our usual places of assemblies, those in the rear somewhat more elevated than those in front, the floor being on a level with the uppermost seat. Over this flight of seats was a thatched roof, twelve feet or so in width, having that part nearest the house, by some means or other, fastened to the lower border of the covering to the main building, while the outer side was supported by eight neatly made posts, placed at distances of eight feet from each other.

Before this royal abode our procession at length made its appearance, having, although not more than a mile distant from the landing, been something like an hour in getting to it, in consequence of several set speeches made by the chiefs, of which we could not comprehend a word, and the arranging, and re-arranging, of the natives in an orderly manner. Marching with all due measure of time, directly towards the front, where, on the flight of seats, attended by at least two hundred of the ladies of her court, sat the queen mother, a large corpulent woman, of about fifty years of age; close by her side was the young king, a very handsome, good looking, stout, and round limbed young man, of about fourteen years of age, possessing a striking and pleasing countenance, open and graceful manners, and an address at once easy, and bespeaking him to be of royal parentage.

The ladies, were all arrayed in snow white cloth garments, with turbans or head-dresses of the same stuff, each one also, had made all due diligence first to besmear her personage with a mixture of the oil of the cocoa and the perfume of sandal wood: but here, the disagreeable odor of the oil destroyed the fragrance of the sandal wood, and had it not been for the presence in which we found ourselves, viz. in the midst of such an assemblage of the king, nobles, and noble ladies of the land, sufficient in itself to command our passive endurance of small grievances, added to which was the respect due to the dames, I do not know how our olfactories could have prevented us from breaking off, and having no further communication with so anti-fragrant a company.

On each row of the seats a place had been studiously kept vacant, except the top one; this was filled from end to end by young females of from fifteen to twenty years of age, apparently. On the uppermost of the three vacant seats, after being introduced to the queen mother and the young king, I was requested to be seated, which of course complied with, by this means being placed between the queen mother and the young king. The next was occupied by the Rev. Mr. Crook, while the remaining, the lowest one, was taken by Toohoorebooa, the regent, who immediately directed a mat to be placed before him and between his feet, for the accommodation of the steward. Directly behind us on the highest seat, sat three young females, to whom the queen mother was pleased to introduce me; giving me to understand, through the interpretation of the Rev. Mr. Crook, that they were her daughters, sisters to his majesty the young king. Upon being introduced, they all answered *Vahvee, Vahvee,* welcome, welcome, accompanying this with a graceful movement of the head and hand. The queen mother then proceeded with the introduction of every one of the other ladies, commencing with the wife of the regent, who was seated at the right hand of our friend Mr. Crook. These were severally called, and as called, introduced; each giving the same smile of welcome, and movement with the head, as the princesses had given. The queen mother acting as spokesman for all the company, making a vast many inquiries as to whether the ladies of my country had plenty of fine cloth, as well as whether they had plenty of good cocoa-nut oil, and the like, to all of which important questions, suitable answers were given, the assembly meanwhile listening with the most profound attention.

An hour or more had been passed in this kind of chat, when, being somewhat thirsty, I ventured to mention my desire for a drink of water to the Rev. Mr. Crook. So soon as the young king understood the purport of this conversation, as explained to him by Mr. Crook, he instantly despatched an attendant in quest of the wished for refreshment. The man soon returned, bringing it in a cocoa-nut shell curiously wrought and figured, which he handed to the young king: after drinking a little, his majesty presented the same to me; upon tasting, I found it to be a tartish mineral water, and therefore handed it to Mr. Crook, with the request that he would try it; after so doing he observed that it was indeed mineral, but the first sample he had seen or tasted, at the same time learning, upon inquiring of the young king, that it had been procured in the mountains; the regent also remarking that it was from a spring up in the mountains, and was much liked by both the queen mother and the young king. To me, however, it was not the thing: I therefore requested Mr. Crook to say to his majesty, that the milk of a young cocoa-nut would be equally as acceptable. Forthwith with, the same attendant, in like manner as at first (bowing as he retired), was directed to bring a young cocoa-nut from one of the trees in front: this was procured without much difficulty, as from the sloping condition in which these trees grew, he was enabled to ascend to the height of between forty and fifty feet, on all fours, that is, on his hands and feet, much in the manner of a squirrel, and with a rapidity that was truly astonishing. After securing the nut he had selected, he descended, bearing it in his teeth by the stem, then husking and cutting off the end, he handed it to his youthful majesty, who, after tasting as he had done with the water, handed the same to me: this I found to be very refreshing.

It had now got to be quite time for us to think about presenting the present which we had prepared for the king. Mr. Crook, therefore, in order that the same might be done in such a manner as to make a suitable impression, spoke to Toohoorebooa, the regent, requesting him to desire all the assembly to remain in silence, as we were anxious to confer a token of our friendship upon their sovereign: he accordingly arose and spoke to the above purport, then reseating himself, all were as still as could be. At this announcement among the company, every eye was turned, first upon the steward, as he busied himself, preparing to pass the piece of plate to the Rev. Mr. Crook. This gentleman, after taking off the paper in which it was wrapped, handed the same to myself. I then arose, and turning to the young king, gently slipped the ribbon over his head, and placed the bright plate upon his breast: Mr. Crook, in a loud tone, at the same time explaining that this was presented as a proof of our friendship; in conclusion, turning and making his respects to the queen mother, who, royal dame, was quite overcome by this great acquisition of wealth for her son, tears following each other as she grasped his hand in her exceeding great joy; yet, not forgetting, however, all the while to move the head and hand in token of her approbation and thankfulness. The three sisters were leaning or hanging during this, on the shoulders of the king, the better to have a good sight of so invaluable an article, and everyone, in ecstasies, was continually crowding around their monarch to see and touch the dazzling present, keeping up the while, women-like, a most astonishing clatter, in evidence, I suppose, of their unbounded admiration for the gift; to complete the scene of confusion, in the exuberance of his joy, the regent chief must needs take me in his arms, give me a glorious hug, and then seat me on his knees: add to this the heat that so great and compact a crowd produced, and it presents a situation that I have no very great wish ever to be caught in again. However, at last content with thus showing off his great love and friendship, the regent let me go, and willingly complied with a request expressed through the Rev. Mr. Crook, to have something like order and silence restored amidst the assemblage, in order that we might speak to their youthful king. To bring about this revolution was not the work of a moment; for some fifteen minutes, or thereabouts, had expired, before it was affected. Gradually the talking subsided into a low humming of many voices, and finally it was sufficiently still for us to proceed with the matter in hand. This we then recommenced, by inquiring of the king if he was pleased to accept the royal badge as a token of our friendship for him. He answered, he readily received this as a mark of our good disposition, which also the regent Toohoorebooa farther confirmed after whispering something to him, and the queen mother, by saying he would answer for the king's accepting it as a token of the greatest value; at the same time exchanging names with me for life, promising that all ships, coming from the nation to which I belonged, whose commanders might hereafter arrive at and visit their harbor, should be treated as their best and dearest friends, as well as supplied with as many hogs and as much bread-fruit as they should wish. In this manner he was going on to explain how he would exert himself to make their reception the more welcome, when farther proceedings were stopped, by an old little dowdy of a woman, attended by a couple of strapping great natives, who came rushing without any respect into the presence of their sovereign, dragging along with them by a rope, a large sow. The animal, it was evident, was of her mistress' way of thinking; for to her squealing, the old woman was adding her continual call, lest we should forget she was there, *"Otee booaugah," "Otee booaugah,"* an expression meaning, "See the fat hog." The clamor made at the presentation of the piece of plate, was a trifle, to this ado.

Directly up in front on the green, came these four animals, the old woman putting the question, without any hesitation, whether I would accept of her sow. Of course, I could do no less than answer in the affirmative; and with equal abruptness she next wished to know should it be sent on board the ship, whether I would give her in return a glass bottle. I told her not to be at all uneasy on this head, for she might safely depend upon receiving the glass bottle, according to agreement. The dame now appeared to be much pleased, and turned away, giving her two attendants orders to lead the sow off; but suddenly recollecting herself, she would first know how I would like to have it sent on board, whether it should be baked first, or sent alive. Not being well enough acquainted with their mode of baking, I chose the last: thus at once terminating a transaction with the only business character I had met with in these parts, and any further speech-making, either by the king, queen mother, or their visitors, this day, at least.

I learned, upon inquiry, that this old woman was no less a personage than the wife of our first and aged friend, Tearoroo, at that time one of the hostages on board our ship, and presuming upon the fact of her being the young king's grandmother, she had in manner as afore described, attended to her own interests, to the complete annoyance and discomfiture of the other affairs. While we were thus employed with, and entertained by, this royal company, a little distance off, yet in open view, at a small palisaded place, one of the natives had been unceasingly, from the time of our coming here, and was till we left, engaged in dancing their friendly dance, usual on such occasions. He was a very handsome strait limbed native, and dressed in a gaudy suit of painted mats, which, as well as his turban, was trimmed off with sundry fine feathers, in order to add to their beauty.

It was now time for us to be thinking about returning to the ship, but before doing this, I desired the Rev. Mr. Crook to present the queen mother and her court, one and all, with an invitation to come and visit the ship the following day. With this she said she was not able, to her great regret, to comply, being, as well as the young king, tabooed from going on the water during this moon, saying, whenever this was expired, she would most gladly come: nevertheless, rather than our visit should be unreturned, her three daughters, together with the other ladies, or at least a large proportion of them, would come on the afternoon of the next day.

We then prepared to take leave of this people, but were first introduced to every individual present. This office the queen mother was pleased to perform herself; nor were the females at all willing to be passed by, or slighted, but each came forward, in regular order, to be presented by the queen mother, and with each we were obliged to touch noses — a most uncourtly sort of ceremony, in my opinion — and one which occupied in its performance at least half an hour. The queen mother and the young king appeared to be greatly affected at our departure, and the later especially, as he held me by the hand, begged that I would stay with them many moons (months). This tedious form of civility at leave taking, which we had very passively endured, being finally concluded, we took up our return march for the ship. The same order of procession as was observed on coming to the assemblage, was now retained in going back.

Seeing the steward carrying my fowling-piece,

which was loaded with some small shot, the regent was curious to know what it was used for, and whether it was a war instrument. In answer to his several inquiries, Mr. Crook informed him, that it could send forth very loud thunder, and also fire, which would kill a native, or any living creature which might be in its way. It very fortunately happened, that just at the moment of our embarking, a flock of small gulls were discovered flying along; and on noticing them, the regent wished to have this power proved upon them: as they came over our heads, I took aim, fired, and very happily succeeded in bringing down one of the number, several of the shot having taken effect. The chief took the bird up, and examined it very carefully, noticing where it was wounded, and the blood as it issued from these wounds, observing, very sagely, as he finished, that "the fire had gone clear through and killed it." This proof of our power I hoped would have a very good and favorable effect, as tending to lead them to have a respect for those who possessed such wondrous means of destroying whatever they wished.

Upon our arrival at the beach, our boat in charge of an officer, was waiting for, and quickly conveyed us on board ship, where the two hostage chiefs had been contented during our absence, and had remained in the cabin all the while. To Tearoroo, the aged chief, I handed the glass bottle, the same being according to agreement made with his spouse, as she had been attentive in fulfilling her part, in payment for the *booaugah* (hog) on board. A little before sunset, two large canoes came alongside, laden with eight large and fat hogs, some pigs, fowls, yams, bread fruit, &c., a joint present from the queen mother and the king. This came totally unexpected by me, and must be considered as an evidence of the firm establishment of the friendly feeling and good understanding subsisting between us, confirming us in our belief, that a favorable impression had been made on their minds; it was also a kind of voucher that our good and highly esteemed friend, the Rev. W. P. Crook, would be well cared for after our departure, as well as such of my fellow-countrymen as should hereafter stop at this harbor for supplies. As an additional proof of their disinterestedness, the regent, who had come along with the present, could not be by any means persuaded to receive anything in return, not even for himself, saying, that by so doing, he was sure he would wound the feelings of the young king and his mother.

The following morning, the ship was surrounded by a number of large and small canoes, in which there were many hundred of the natives. With these, and several chiefs and attendants, on the ship's deck, we were doing a pretty brisk trade, as a plunge was heard, followed by the sentry at the stern calling out, that a native had jumped out of the cabin window, having something with him which he very carefully held to his breast by one arm. We immediately mustered at our quarters, and an officer was directed to ascertain what was missing. This was found to be the azimuth compass, which had been left standing on the after locker, near the windows. The chiefs and natives now began to be alarmed, and commenced leaving the ship: the regent, too, happened to be on board at the time, and while the Rev. Mr. Crook made known our loss to him, and the importance the article thus taken was to us, also informed him, that neither himself, or the chief his brother, could be permitted to leave the ship, however unpleasant and painful such proceedings might be, until the compass was returned. He answered, it should be brought back, and for this purpose despatched an under chief, one of his attendants, to the shore after it. Thinking, meanwhile, that it would be good policy to show forth the effect produced by our fire-arms, I requested Mr. Crook to call the regent's attention to the quarter-board, which was perfectly sound, then taking a small pocket-pistol, loaded with a single ball, I discharged it, the ball of course making a hole through the board. This, he was given to understand, would have been the case had it been shot at a man. Mr. Crook, as he mentioned this, also calling to his recollection the gull which he had so closely inspected the evening previous; yet that they might see whether the captain could destroy them, together with their village, and even their island, he would fire off one of the big guns. Accordingly, after drawing the shot from one of the carriage guns, the piece was pointed towards the shore, and discharged. The report considerably alarmed the natives and for some little time they were the very pictures of fear and terror, listening very attentively to the echo as it was repeated from the mountains; the regent alone venturing to observe, that he heard it break the rocks and tear up the trees on the mountains. This remark increased their alarm to such a degree, as to cause them to ask Mr. Crook to beg the captain would not destroy their valley; but knowing that he would understand me, I thought proper to give an evasive answer to this request, supposing that to appease me they would the sooner return the lost compass. Every islander now left the vessel, with the exception of the two captive chiefs and their immediate attendants. One of these being also a chief, the regent requested and obtained permission to send him on shore in one of his remaining canoes, in order if possible to hasten the returning of the compass. An hour was passed in this state of alternate hoping and doubting, before the two chiefs came back with the stolen article; it was broken into a number of pieces, as was also the glass and box; after some time and labor spent in matching the pieces and putting them together, all was found except the vane; this was told the regent, and he instantly made inquiries of his messengers concerning it, from which it appeared, that the native who had taken the vane, had gone a long way up the mountain, but that several had been sent after him, and the king, added the regent, has pledged his word, and so do I mine, that it shall be returned in the morning.

Every effort appeared to have been made, that lay in their power, towards restoring our lost compass, by both the king and his chieftains; I thought it was therefore time to renew the good understanding which had been thus temporarily broken in upon, and as there was no remedy for our loss, the natives in their ignorance having broken the compass to pieces, we were obliged to make the best of it as it was; taking the regent and his brother, therefore, by the hands, I expressed a hope that now our friendship and confidence in each other, if possible, would be strengthened, inasmuch as their exertions to secure and restore our stolen property, clearly proved that they were not at all concerned in, or had even countenanced the theft, but which we now found had been committed by one of the bad men of their nation: although our friendship had been disturbed by the occurrence, yet I requested him to say to the king, that my attachment for his majesty was increased by the interest he had taken in the affair. All this pleased the regent, and early the next morning he came again on board, bringing along the missing vane, saying, he had caught the thief, and had him in safe keeping, wishing to know whether I would not like to have him knocked in the head at once; this was a mode somewhat more severe and expeditious for punishing the poor fellow, than I cared to follow. I told him no, but added, that if he or any of his friends, the chiefs, should see the thief on board or near the ship again, and would let me know it, I expected then to be able to punish him more according to my own notion. This ended the matter and restored harmony and confidence again between all parties. Again the natives came on board, and again trade was in full operation.

At eleven o'clock this day, the chiefs with Mr. Crook left the ship and went on shore to assist the ladies of the land in paying their appointed visit. The better to honor said visit, our colors were all set, previous to their going away. Somewhere about two o'clock in the afternoon, which was the appointed hour,— nor were these ladies so long in making their toilets as those of my native land, making a reasonable comparison between the two operations of lacing, &c., and their rubbing on an extra quantity of cocoa oil for the occasion, for within five minutes of the hour appointed, their escort found them ready, — a rare thing at home. Our boat was sent to lend a hand, if needful, but this was not the case, as several of their largest canoes were shortly to be seen paddling on towards us, and in a few minutes, fifty of the dames of the court were, without much difficulty, got on board, all duly oiled and perfumed, and suitably attired in their snow white muslin-like dresses, and turbans or head ornaments, attended too by a goodly proportion of chiefs, Toohoorebooa, the regent chief, at the head. Everything that might be a temptation, had been removed to a secret place, and those articles of a more portable nature, were placed under lock and key, so that we were enabled to give free permission to visit and inspect all parts of the ship, without any apprehension of being the losers by our courtesy. The wife of the regent performing a like office for the females, which the husband did for the men, a sort of director-general, took the lead, attended by the three sisters of the king, their conduct throughout being very decorous, frequently stopping to know the names of different things, their uses, and all about them. I embraced the opportunity thus offered, for making an arrangement with the three princesses, and agreeing with them for three dresses and turbans like those they wore on this visit; Mr. Crook, on their return to the shore, showing how to fasten them with pins, so that their shape and form could be retained until they were delivered into the hands of our ladies in New

York, which we did on our return. The idea pleased these women very much, and among other of their inquiries, they wished to know if our ladies would walk about with them, and wear them in company, and afterwards make many more like them. (This cloth is made from the bark of a tree, and by being stained, is of a variety of colors; yellow and white are the most prominent, chiefly the latter). We paid every attention, and endeavored to please these good hearted people, by all the means in our power, and when they had finished viewing the various arrangements of the ship, with many of which they were greatly astonished, we gave a finish to the whole, by displaying what in their estimation was our riches, a general assortment of such articles as ribbons, looking-glasses, buttons, beads, scissors, needles, &c.; the general exclamation on beholding these, being, *Motakee! etee Motakee!* "Good, see, very good! or beautiful."

By the assistance of our excellent friend Mr. Crook, a present had been prepared for, and was now presented to each of the females, the three sisters taking with them one for their mother, the whole party going off, after this, apparently very much delighted, to their homes.

Having mentioned to the Rev. Mr. Crook that my intention on the following day was to trade for provisions, hogs, &c., and the day after that to sail, I at the same time made a tender of my services, provided he had fully determined to remain behind, to take any letters which he wished to send home. He accordingly, having pen, ink, paper, and every convenience at his service, prepared, and entrusted to my care, a packet of letters for the Society in London. This packet, on our arrival at New York, was forwarded to its destination.

The report of our intended departure in a day or two, as mentioned by Mr. Crook on his return to the shore, spread rapidly among the islanders; the result of which was, that the next day there was more business than sufficed us. Towards sundown, the regent chief, Tearoroo, in company, came alongside in their canoe, bringing several fat hogs, fowls, &c., the same being a present from the young king. In return for this, and to each of these chiefs, for whom I could feel nothing less than the strongest attachment, I made up an assortment of axes, hatchets, knives, and fishhooks, which was fully satisfactory to them.

With my esteemed friend, the Reverend gentleman, I then divided my wardrobe; and the better to insure to this truly Christian man, the respect of the natives, added a fowling-piece, with a suitable stock of ammunition, as also a lot of axes, cutlery, toys, &c. for trading purposes; a secretary, paper, quills, ink, penknives, and all which would tend to assist him in discharging his important duties, while at the same time it would make his leisure hours pass pleasantly along.

May 29th, 1798. Made the signal for sailing from the harbor of Paypayachee (so called by the natives). This was observed by the Rev. Mr. Crook, and our friends the chiefs, who had been on board the afternoon previous, and before we could get under way, they were on board to make a parting visit; the regent saying the king was very desirous that we would remain with them one moon longer, and if this was too

A WARRIOR OF THE MARQUESAS ISLANDS
From a photograph made about 1900

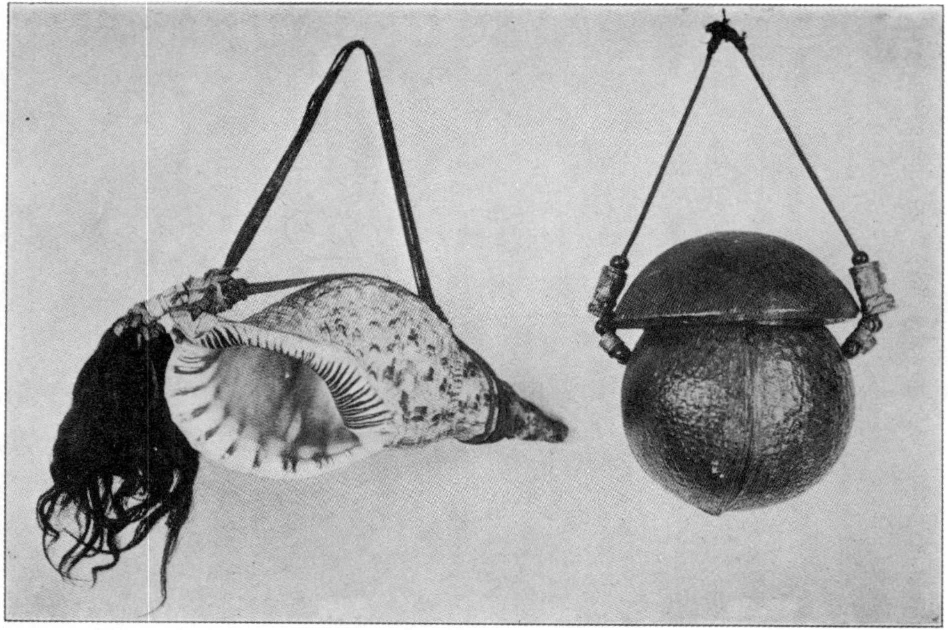

WAR CONCH AND FOOD BOWL, MARQUESAS ISLANDS

From a photograph of the originals now in the Peabody Museum, Salem.

a. War conch (Triton shell) ornamented with human hair and bones, brought to Salem before 1821.
b. Cocoanut shell food bowl with wooden cover and human bone ornaments, brought to Salem before 1802.

long, that we would not go for at least five suns. "The king," said he, "has got plenty of hogs and bread-fruit, plenty for you, and for all his people." There is nothing on earth so difficult in its performance, or that requires so much care and caution, as to refuse, without giving pain, the offer or invitation of a generous heart — nothing so distressing, to such an individual, as to refuse to accept what he offers from the purest and most disinterested motives; and that these were the actuating principles of the offerers in this matter, I cannot doubt: therefore the difficulty before me, for an apology the most likely to answer my purpose. I replied, after expressing many thanks for his majesty's invitation, that the order for sailing had been given, it was imperative, and like their taboo, must be obeyed, and the ship sail this very day. It was not satisfactory, however; the end was not attained. Dejected and disappointed, the regent sat down, and without uttering a word, rested his head upon his hands some time. I knew not exactly what to do, but with a view to relieve his mind, I told him that perhaps we might return to their harbor again in twelve or eighteen moons, but he ought to let us go in a cheerful manner, and not be so dispirited: there he still sat. Meanwhile, at eleven o'clock, we had taken up our anchor and stood out of the harbor; having passed a mile to seaboard, from and without the two islets which we called the sisters, the ship was hove to, so that those friends who had remained on board thus far to see us safe out of the harbor, might get into their canoes and return.

From the aged chief Tearoroo, I received a friendly

hug, and his good wishes, after having "touched noses" with him. The parting with my excellent and worthy friend, the Rev. Mr. Crook, was very impressive; but from Toohoorebooa, the regent, there was no getting away; several times he bade me farewell, would get as far as the ship's side, and then return, seize my hand, and beg me not to go yet, not so soon. To try the advantage of a little coaxing, some axes, and articles that he was known to be partial to, were brought upon deck; of these he was requested to make choice, or if he wished it, to take them all, and whatever he chose should be put into the canoe for him. He did not even seem to notice what had been said, but stood still, as one altogether indifferent to what was taking place around him.

Mr. Crook, and Tearoroo, who had patiently waited for a long time, supposing that if the regent should find he was likely to be left without a canoe in which to return to the shore, would bestir himself to take advantage of theirs, recommended me to fill away on the ship, and cast off their rope; this was done, the canoe keeping within a few yards astern. I finally persuaded this warm-hearted chief, so much as to get him on the outside of the ship's railing, where he remained some moments, standing on the main channel, in a thoughtful mood. "Well," said he, at last, "let me then bid you 'good bye' once more;" and after so doing, according to their custom, he let himself fall into the sea, making the best of his way for the canoe, which, as he reached it in safety, the three friends paddled for the shore. At eighty yards distance, the regent arose, and gave us a friendly wave of the hand;

and so long as we could see him, with the aid of the glass, continued with his head resting on his hands. He was somewhat over the middle stature, full six feet in height, of a straight and handsome form, a man possessing immense strength, of a noble disposition, ever ready, even at his own inconvenience, to do all in his power for the accommodation of his visitors; and had it not been for the disfiguration of his face and ears caused by the tatooing, with which ornament (according to their notions of beauty), or mark of merit, he was much more distinguished than any other native we saw, he would have been a good subject for the study of the sculptor and the painter unitedly; but as it was, the effect of this tatooing was to make his skin appear as dark as that of any African. This custom is not confined to the men, for the women likewise indulge in it in a greater or less degree, the same being held in very great estimation, and considered as a mark of the highest honor.

Among the men, there was a prevalent fashion of cutting the hair off close to the scalp, as far as the middle of the head, leaving on each side a long lock, which was either left hanging loosely down, or bound up in a bunch on the crown, according to the pleasure of the individual. The women appeared to be divided into two parties on this matter: the better class wearing their hair without any kind of fixing about it, down the back, or partly over the shoulders in front, and also down the back, negligently; while the lower class always cut theirs short off.

Both the men and women were habited alike, the mildness of their climate not requiring much clothing; it generally consisted simply of a covering around the waist, yet at times a more substantial garment was used.

The men are all well formed, and are exceedingly active, either on land or in the water; and the women, were it not for their color, which is of a light copper cast, would be considered as beautiful, when arrayed in bleached white robes and turbans, their long black hair, which they comb as carefully as do our ladies theirs, hanging loosely around; their appearance at least is such. Both men and women have excellent teeth, as white and sound as one could desire to possess.

They are not so numerous, or warlike, as are the inhabitants of the Island of Wepoo, and although at times engaged in wars, particularly with the tribe situated to the eastward of them, yet in the main they appeared to be a peaceable friendly people; this had been their bearing towards ourselves, and the universality of this disposition, especially with the chiefs, evinced during our stay among them, certainly established their claim to our grateful remembrance.

The harbor of Paypayachee is excellent, and possesses many conveniences: it is situated in a conspicuous bay, at the side of which, and abreast of the beach, was our anchoring station. In searching and sounding we did not discover any rocks, or dangers under the surface of the water, so that a ship may stand boldly in for the sisters (a good mark to know the harbor by), when these are passed, hauling up towards the beach on the east side, and anchor at such distances from the shore as suits, with the water varying in depths of from ten to twenty fathoms, according to the distance from the beach. If the trade blows far from the northward of east, a large ship had better anchor near the sisters, in the passage, and warp in, as the wind from the mountains will then be baffling in the bay with variable flaws, strong puffs, and with calms, yet a small vessel, or fore and aft schooner, can run in even at such times, with perfect safety.

There are other harbors, as I was informed by the chiefs, through Mr. Crook's interpretation, viz. two to the eastward, and one to the westward, on the south coast; and also two on the north coast of this island.

The valley which is called Tiuhoy, is in form not unlike a triangle, being a league or so wide, near the shore, gradually narrowing as it goes in shore, where at its farthest end are a number of trees, while on the two sides are high mountains, being, with the exception of now and then a ledge on which are the dwellings of the islanders, almost a complete perpendicular wall. As seen immediately after a shower, when the sun is up and shining upon the numerous streams which then pour down the sides of the mountains, and over the green of the valley, looking like so many layers of polished silver; with here and there a native's habitation, some built very high up the mountain, like so many bird's nests, the valley then presents one of the most beautiful and romantic views mortal eye ever beheld.

This tribe is called the Taeehs, the neighboring tribe in the valley to the east of them, Happah, and that to westward the Tiohahs. At the period of our arrival, hogs, fowls, bread-fruit, cocoa-nuts, yams, and tarro, were very plenty.

Their sugar cane I certainly think is the finest as well as the largest that is grown. Refreshments we obtained at most reasonable rates: a pig, for instance, weighing forty or sixty pounds, was purchased with a small fish hook, board-nail, or a bit of iron hoop three or four inches in length; other articles at a like cheap rate. In the workmanship of their war-clubs, spears, clubs, and paddles, their canoes, as well as their mother-of-pearl fish hooks, and gear, which are all surprisingly neat, and extremely curious, considering they were not made with steel or iron tools, but by their rough implements of stone, shell, or bone, they have shown themselves to be first-rate natural mechanics. We purchased many fathoms of their fishing lines and ropes, made from the bark of a tree, both of which were very neat articles; the last we found answered as running rigging very well indeed. Their war arms consist of the spear, the war-club, and the sling. Their martial music is such as is produced by the drum (a hollow log with a sharkskin drawn over each end), and the war conk, a large shell of various colors, the whole curiously polished and ornamented with human hair, either braided or in bunches.

They too were cannibals, or at least so far as the eating of the flesh of their enemies makes them such; as a proof of this I observed one day while trading with the canoes alongside, something wrapped up in some palm leaves, on board of one of them, the native in which did not offer it for barter. This was so unusual, that I examined it, and found the same to be a piece of human flesh, baked; surprised, I shrunk back with horror, and asked him what he was going to do with it. The fellow took my meaning, and replaced the leaves around it as formerly, only answering, that as it was a part of one of their enemies, it was therefore very good food for him, and whenever he was hungry, he was going to eat it. This, was the only evidence that came under my immediate observation: I endeavored to make him comprehend how wicked and awfully disgusting such a practice was; with what success, however, I know not.

When our ship first came in sight of the Marquesas Islands, the crew had been gathered together, in order to hear and know a set of rules and regulations, that had been prepared for our government in all our future dealings and trade with the natives; the penalty for a breach of these, was immediate confinement in the ship's run, until such times as she should leave the island or place where the fault was committed; or if the crime did not warrant so severe a punishment, one more proportioned to it was to be inflicted. These were highly necessary, for the prevention of mutual imposition; the better to insure attention to them, as well as to prevent any one from having an excuse for doing wrong, and to superintend the purchasing and selling, an officer was daily appointed. He superintended all the business of bartering and trading, and inspected all the purchases received on board, of whatever kind they might be, while any person wishing to possess an article of curiosity, without having the means so to do, had them charged to their account on the ship's books.

To the credit of our crew, there was but one breach of these regulations; and this was with a young man who had had in his chest an old fish hook, which was noticed by an islander near by, who offered to buy the same with one of their curiously fashioned fish hooks; this was accepted, and the bargain closed. Shortly after, the native being wanted, was called aft by his chief, and appeared with the hook in his lips. When asked where he obtained it, he was confused; the chief then required him to point out from whom it was had, when the sailor, apparently just sensible of his fault, acknowledged it, and was sent down into the run. His confinement, however, was not very lengthy, for his shipmates all came aft with a petition for his release, whereupon, in consideration that it was a first offence, after the customary promises of good behavior, better care in future, &c., on his part, he was returned to duty.

CHAPTER XII

NUGGOHEEVA TO CANTON

MAY 30th, 1798. After the departure of our three friends, as related in the preceding chapter, we bore up and stood to the westward along the coast, going off before the wind with all sail set. At 6 P. M. the north-west point of Nuggoheeva bore E. by N. about five leagues distant; the Island of Fettooeeva then bearing N. W. by W. six leagues distant. It was now but five or six days since we had first obtained sight of the Washington group of islands, during which, a harbor had been searched for, and found, water for a long passage had been filled, and an abundant stock of excellent refreshments laid in, of which the present appearance of our decks gave ample testimony; so that the time spent at these most beautiful islands, cannot in anywise be said to have been foolishly squandered.

Among our crew there were some mechanics, all, with one exception, first-rate workmen; assisted by the ingenuity and perseverance of these men, we had run a hand-rail, with a straightened sheer fore and aft the ship, about a man's height from the deck, the space between it and the main rail being filled up with diamond mesh net, faced with canvas. This, the better to agree with the bulwarks, was painted in a like manner; so that when our boarding netting was down, and stowed in form on this rail, it had the exact appearance, at a trifling distance, of hammocks laid up in the netting. Our carpenter's mate having served at the turning business, also contrived to furnish himself a lathe and set of turning tools; with these he had made some quakers (wooden, or false guns), which, duly provided by the carpenter and armorer with carriages, vents, tompions, aprons, breechings, all in complete style, and being well painted, were such good imitations of the iron guns, as to require a close inspection to detect the deception. This may be better conceived from the fact, that when sold with the ship on her return to New York, the purchaser, although he had been on board daily, while the cargo was being discharged, and had stood upon them very frequently, bought them all as so many iron guns, but when discovered to him, came with a complaint, asserting that he had been imposed upon: he looked, as we seamen say, "Like one struck all aback," when informed that the inventory merely stated, armed with carriage guns, without specifying the exact number.

So that by these means our little ship was as to her outward appearance, completely changed from a merchantman into a man-of-war. The individual excepted as above, although entirely ignorant of the business for which he shipped, and knowing nothing of even the easiest branches of his trade, as a cooper, was nevertheless a good-natured fellow, and being willing to learn, was at this time set to work to bear his part in the various duties on shipboard, at making buckets, kegs, cans, kids, &c., in which he shortly became quite an adept.

At 5 A. M. the lookout at the mast-head, gave notice of his seeing land in the N. E. This, as we neared it, appeared to be two islands, one high and of considerable extent, and the other a low island. The first we named New York Island, and the second Nexsen Island, in compliment to my friend and owner Mr. Elias Nexsen. The number of smokes that could be seen arising in various parts of the land, we considered as indications that they were inhabited, but bearing N. E. by E. from us, at the distance of from four to six leagues, with the wind from the E. N. E. did not think proper, particularly as we were not in want of anything from them, to waste time in beating up to them. At noon, our latitude was 8° 13′ south, longitude 141° 31′ west; at which time the middle of New York Island bore E. by N. eight leagues distant.

June 7th. The trade wind, with which we have been favored for the last few days, has blown a moderate breeze during the same, varying from the E. to the N. N. E. with occasionally, heavy showers of rain. Boobies, man-of-war hawks, small white gulls, tern, or egg-birds, in great numbers, always in company, together with many flying-fish, of a larger size than ordinary; these were a favorite fish with us, and came very opportunely just now, by adding to the variety of our dishes. At 4 A. M. crossed the equator for the second time, being in longitude 154° 43′ west.

June 11th. Had the trade wind light from the E. S. E. with passing clouds; at intervals light showers of rain. At 3 A. M. soon after one of these showers, the seamen at the masthead on the lookout, gave the cry of "Land ho!" and in reply to the demand of the officer, then in charge, of "Where away?" replied, "Direct ahead, and close aboard;" the ship under full sail, with steering sails set below and aloft, was at this time going before the wind. The orders for the officers having charge of the deck, while sailing across this extensive ocean, were, in case land, or any dangers should be discovered during the night time, for them not to leave the deck to report the same to myself, but to give their immediate attention to the helm and the sails, have all hands called, while at the same time by stamping on the deck, over my head, I should be awaked. In the present instance, these directions had been promptly complied with, by the officer then in charge, so that in a few seconds I reached the deck, just at the moment that the lookout again called out, "Breakers close aboard;" the helm was then a lee, at port, and the ship coming fast to the wind on her starboard tacks, the studding sails were coming in, and yards bracing up, when stepping to the larboard rail, the land was seen stretching along in a direction seemingly about north and south, with the surf in the western board, as a rain shower passed over, and its clearing up enabled us to see it, appearing to be one continued sheet of white foam along the horizon, breaking high, with a tremendous noise, on the coral reef that bound the coast, and about one mile distant from the ship, as seen from the deck. After trimming every sail upon the wind, the ship looked about two points off the land, moving to the north-east at the rate of two and a half miles per hour, a heavy swell or sea from the eastward heaving her to the leeward, as she ranged along the land.

After an hour passed in great anxiety, a point or

cape was discovered off the fore-beam, while as the day began to dawn upon us, it was observed that we had fallen in with the land, about four miles to the south of the north-east Cape,* from which we were now gaining an offing to the north, and whence the land tended away to the westward. The ship had not gained over two miles to the northward of this cape, when it became suddenly calm, and by 8 A. M. the leeward swell or sea had hove her much more than a mile along the coast to the westward of this cape.

In this instance, we narrowly escaped shipwreck, as on sounding with the boat, for the purpose of ascertaining if such would probably have been the case, we were led to infer as much from our not being able, with fifty fathoms of line, to obtain bottom at a cable's length from the coral reef which binds the shore; so that had the calm happened only two hours sooner, it would unquestionably have taken place. At 10 A. M. a sea breeze from the north-east sprang up, whereupon we bore away, and sailed along the northern coast to the westward: this island appeared to be one of some extent, and the group to consist of three, all within sight at the same time, and laying in such position to each other, as in some measure to form a triangle. The north and south islands were each about nine miles in length; the other, the easternmost one, stretching to the northward and southward, and adjacent to

*Which we named Cape Brintnall (after our meritorious first officer, now Captain Brintnall, a gentleman of much nautical information, residing in the city of New Haven, and who, subsequently to this, has been in command of a ship for many voyages, in the business of seal fishery and commercial trade to the Pacifics and Canton, and while thus engaged, has paid for duties on China goods, many thousand dollars, the result of his great industry and perseverance, into the United States treasury).

the eastern ends of the two first mentioned, was about six miles in length, the whole three forming a most spacious bay, with good anchorage and good harbors. At noon, being off the north-west port, we hauled in under easy sail, over a bank which lays off the western end of the islands, where a ship, abreast of a passage into the bay, may anchor under their lee.

On gaining this position we hove to, hoisted out the boat, manned her, and pulled up through the passage. The landing we found perfectly smooth, and effected by resting the bows of the boat on a small sandy beach, at the starboard hand, as we passed into the bay. On the south island, and near by a grove of cocoa-nut trees, whose fruit then lay strewed around, covering the ground from one to three feet deep, and seemed to have ripened and thus fallen for many years past, our boat's crew, having formed themselves in a line from these to the boat, very quickly loaded her from the upper course of those nuts which had fallen last, by passing them briskly from one to another; meanwhile, I employed myself in taking a kind of fish, much like the striped bass. Of these there were great quantities continually crowding against the boat, so that it was an easy matter to spear and take them, without letting the shaft of the grains go out of the hand. After getting upwards of fifty, weighing from five to twelve pounds each, I desisted, supposing that this number would be full as many as we could consume on board ship before they should spoil: when cooked, they were found to be very finely flavored, and good eating. The sharks here are very numerous, and while the boat was on her passage into the bay, before she entered the pass, they became so exceedingly ravenous around her, and so voracious withal, as frequently to dart at, and seize upon her rudder and the oars, leaving thereon many marks of their sharp teeth and powerful jaws; but so soon as she left the pass and entered within the bay, they deserted her, their stations being instantly occupied by multitudes of fish, less rapacious, yet infinitely more valuable. When the boat was loaded, accompanied by an officer, the steward going along, we took a stroll into the interior for a few minutes, among the upland grass and groves of various kinds of trees, without being able to discover any of the valuable bread-fruit tree. At the barren spots, the birds, boobies, knoddies, and the like, were quietly setting on their nests, so fearless and gentle, as to be easily taken by the hand; yet in self defense, sometimes pecking sufficiently hard to draw blood. Amongst the birds, was one species about the size of our robin; with a breast of scarlet colored feathers, the under portion of the body being finished off with bright red, the neck of a golden color, back a lively green, with a yellow beak, except the very points, which were of a light dun color, the wings and tail being both of a jet black, and the last tipped off with white: it was a most beautiful and lovely bird, with its brilliant and richly variegated plumage. We were much chagrined, while observing these, to see a man-of-war hawk flying by with one in his mouth, apparently having just caught it. At 6 P. M. returned to the ship, with the result of our afternoon's operations.

There were no signs nor vestige of habitation discovered by us during our perambulations, from which we could infer that mortal ever had placed his foot upon these shores, previous to the date of our arrival; still Captain Donald Mackay, in a vessel under the agency of the author, a few years after their discovery, being at anchor some weeks at Fanning's Island, while procuring a cargo of beach la mer, turtle shell, &c. for the China market, reported on his return home, that during this stay he frequently walked into the interior, and in one of these walks had come across some heaps of stones, which, to all appearance, from their order and regularity, were thus placed by the hands of men, although from the coat or crust of weather moss with which they are covered, it must have been at some very remote date. Being prompted by curiosity, and a desire for further information upon this subject, he caused one of these piles to be removed, and found it to contain, a foot or two under the surface of the ground, a stone case, filled with ashes, fragments of human bones, stone, shell, and bone tools, various ornaments, spear and arrow heads of bone and stone, &c.

These islands are situated in latitude 3° 51' 30" north, longitude 159° 12' 30" west, and as before stated, are three in number, exclusive of the islets. We gave them the name of Fanning's Islands, and by this they have been recorded, and remain on the charts in use. There is sufficient depth of water through the passage for any merchant ship to pass in, and on the inner or bay side is smooth and convenient anchoring, which, together with the abundance of wood and water, the tropical fruits, best of fresh fish, and excellent turtle, here to be obtained, make this a very desirable place, for the refitting of a ship, and refreshing a crew. The soil, generally speaking, as it appeared throughout the interior, was rich and luxuriant. The anchoring, on the bank off the western ends, and under the lee of the islands, is from twenty to thirty fathoms, over a sandy bottom: this, as the trade winds here prevail, will be found to be a smooth and easy roadstead; in the ebb and flow of the tide, the current runs in and out of this bay.

Having recruited our stock of fire wood, as well as a goodly quantity of cocoa-nuts, at 7 P. M. we weighed anchor, and sailed from Fanning's Islands, then steered under a moderate trade breeze, to the N. W. by W. the weather fair and pleasant, attended by a smooth swell of the sea from the eastward.

A little before noon, June 12th, the seaman at the mast-head again called out, "Land ho!" adding, that the same was half a point off the lee-bow. At meridian, this newly discovered island bore west by north four leagues distance. This, was of a much greater elevation than Fanning's Island, and was, moreover, covered with plants or grass, presenting to our eyes a beautiful, green, and flourishing appearance. With the unanimous approbation of every individual on board, both officers and seamen, and with feelings of pride for our country, we named this, Washington Island, after President Washington, the father of his country.

Having but recently obtained a bounteous supply of refreshments, there was no necessity for our making a landing here, although the trees and green foliage, among which we plainly saw the tall cocoa-nut tree, presented a very strong inducement for us so to do, but passed it to the south, we then steering to the west. At 2 P. M. our ship was abreast of the island, having it between one and two miles distant, off the starboard beam. The waters now, were filled with a vast many fish of different kinds: of these we caught with the grains several excellent well-fed fellows, much resembling the king fish taken in the West India seas.

There can be no doubt but that at this island a vessel might obtain an abundant supply of excellent refreshments for her crew. As at Fanning's Islands, so here, we could perceive no tokens of its being at all inhabited. Washington Island is in latitude 4° 45' north, longitude 160° 8' west, and lies in a N. W. by W. direction from Fanning's Island, at a distance of some twenty-seven leagues. As we passed it, we discovered a coral or sand bank off its western side, extending a mile and a half from the shore, where it appeared a ship might come to anchor: from the south-west point, a coral reef, on which the sea breaks, puts out into the sea about a quarter of a mile. We left this beautiful island under a whole sail breeze, and at 6 P. M. it bore E. S. E. at a distance of about six leagues. From the many flocks of birds hovering over and around us at this time, particularly a small dark brown bird, with a white crown, which had not before been seen so far from the land, we were inclined to think that still more land existed in our vicinity; yet were not able to discover any other, notwithstanding we had remarkably fine weather, and kept constantly a sharp lookout aloft for the purpose.

June 14th. Although somewhat foggy around the horizon, yet we had the weather quite pleasant, with a brisk trade breeze, nor has there been any necessity, while sailing over or across the western part of this extensive Pacific ocean, to lay the ship by a single night, through fear of running her upon any hidden danger, the weather having been remarkably fine all the time, with moderate trade winds, ever keeping a good lookout, and believing ourselves perfectly secure from this precaution; as usual attended by a great many of the feathered race, our constant companions. In this manner prosecuting our voyage, it seemed more like a sailing excursion, or party of pleasure, at least this portion of it, than what it in fact was.

The following occurrence, although bordering upon, and seemingly partaking of the miraculous, did nevertheless, actually take place. At nine o'clock in the evening, my customary hour for retiring, I had as usual repaired to my berth, enjoying perfect good health, but between the hours of nine and ten found myself, without being sensible of any movement or exertion in getting there, on the upper steps of the companion-way. I suddenly awoke, and after exchanging a few words with the commanding officer, who was walking the deck, returned to my berth, thinking how strange it was, for I never before had walked in my sleep. Again I was occupying the same position to the great surprise of the officer (not more so than to myself), after having slept some twenty minutes or the like: here, upon observing the glittering stars overhead, and feeling the night air, I was preparing to return to the cabin, after answering in the affirmative his inquiry whether Captain Fanning was well; why, or what it was, that had thus brought me twice to the companion-way, I was quite unable to tell, but lest there should be any portion of vigilance unobserved by those then in charge, I inquired how far he was able to see around the ship; he replied, that although a little hazy, he thought he could distinctly see land or danger a mile or two, adding, that the lookout was regularly relieved every half hour, in reply to my question if such was the case. There was something very singular in all this, and with a strange sensation upon my mind, after what had passed, I again returned to my berth. What was my astonishment on finding myself the third time in the same place! yet with this addition: I had now, without being aware of it, put on my outer garments, and hat; it was then I conceived some danger was nigh at hand, and determined me upon laying the ship to for the night; she was at this time under full sail, going at the rate of five or six miles per hour; all her light sails were accordingly taken in, the topsails were single reefed, and the ship brought to forthwith on a wind: leaving directions with the officer in charge to tack every hour, and the same to be passed to the officer who should relieve him at twelve o'clock, so that by these means we might maintain our present station as near as possible until morning; I added a request that he would call me at daylight, since himself would then be on the watch. He was surprised, and looking at me in astonishment, appeared half to hesitate whether to obey, in all probability supposing me to be out of my mind. I observed to him, however, that I was perfectly well, and possessed of my right senses, but that something, what it was I could not tell, required that these precautionary measures should be studiously observed. After leaving these necessary directions, a few minutes before eleven I once more retired, and remained undisturbed, enjoying a sound sleep, until called at daylight by the officer. He reported the weather then to be, as it had been during the night, much the same as the evening previous, with a fine trade wind from E. N. E. Giving him directions to keep the ship off her course and make all sail, after attending to some little duties, I followed to the deck just as the topmost rays of the sun came peering above a clear eastern horizon.

The officers and watch were busily engaged in the washing of decks, and attending to those various duties which claim attention at this portion of the twenty-four hours. All was activity and bustle, except with the helmsman: even the man on the lookout was for a moment called from his especial charge, and was then engaged in reeving and sending down on deck the steering sail haulyards. This induced me to walk, after taking a few turns back and forth on the weather side of the quarter-deck, over to the lee quarter, not expecting, however, to make any discovery, but solely to take a look ahead; in a moment the whole truth flashed before my eyes, as I caught sight of breakers, mast high, directly ahead, and towards which our ship was fast sailing.

Instantly the helm was put a lee, the yards all braced up, and sails trimmed by the wind, as the man aloft, in a stentorian voice, called out, "Breakers! breakers ahead!" This was a sufficient response to the inquiring look of the officer, as perceiving the manœuvre, without being aware of the cause, he had gazed upon me to find if or no I was crazed; now, however, casting a look at the foaming breakers, his face, from a flush of red, had assumed a death-like paleness. Still no man spake: all was silence, except the needed orders as promptly executed, every man moved, and every operation was performed in the manner, and with the precision that necessarily attends the conduct of an orderly and correct crew.

The ship was now sailing on the wind, and the roaring of the herculean breakers under her lee, at a short mile's distance, was distinctly heard, as the officer to whom the events of the past night were familiar, came aft to me, and with the voice and look of a man deeply impressed with some solemn convictions, said, "Surely, Sir, Providence has a care over us, and has kindly directed us again in the road of safety. I cannot speak my feelings, for it seems to me, after what has passed during the night and now also appears before my eyes, as if I had just awaked in another world." "Why, Sir," continued he, "half an hour's farther run from where we lay by in the night, would have cast us on that fatal spot, where we must all certainly have been lost. If we have, because of the morning haze around the horizon, got so near this appalling danger in broad daylight, what, Sir, but the hand of Providence, has kept us clear of it through the night." With him I perfectly agreed, and answered, that we should now be truly thankful to that Heavenly and protecting Being. But urgent and imperative calls for attention to our perilous situation, forbade at present any farther remarks; the officer forthwith took the glass, and went aloft for the purpose of ascertaining whether the ship was nearest to the north or south end of the reef, as also whether we were likely to weather and clear it on this tack.

I freely confess, that this premonition, so unusual, and the transactions therewith connected, are deeply impressed upon my mind, as an evidence of the Divine superintendence, and there, ever will remain so firmly imprinted, — how could they otherwise be? — as never to be erased. All hands, by this time made acquainted with the discovery, and the danger they had so narrowly escaped, were gathered on deck; gazing upon the breakers with serious yet thoughtful countenances. We were so fortunate as to weather the breakers on our stretch to the north, and had a fair view and overlook of them from aloft. It was a coral reef or shoal, in the form of a crescent, about six leagues in extent from north to south; under its lee, and within the compass of the crescent, there appeared to be white and shoal water. We did not discover a foot of ground, rock, or sand, above water, where a boat might have been hauled up; of course had our ship run on it in the night, there can be no question but we should all have perished.

Bearing away again on our course around its northern side, I went aloft, and with the aid of the glass could plainly see the land over it, far in the south; yet regretting very much that we were not able to spare time to make a lengthy visit, the better to ascertain its extent and productions; but this we could not, in justice to all interested, undertake. The shoal is situated in latitude 6° 15′ north, and longitude 162° 18′ west.

Captain D. Mackay, formerly spoken of, in command of the schooner *Brothers*, under the agency of the author, visited this land, which has received the name of Palmyra's Island, a few years after our discovery of it, while prosecuting his voyage in search of beach la mer, turtle shell, &c., for the Canton market, and has given the following particulars thereof. "Palmyra's Island lies in latitude 5° 49′ north, longitude 162° 23′ west, is about three leagues in extent, and has two lagoons on it, in the westermost of which, there is twenty fathoms of water over a coral and sandy bottom: the approach to the western part of the island is rendered dangerous on account of the coral rocks, just below the surface of the water, which extend out from the shore to the distance of three leagues; but on the northwest side there is anchorage three-quarters of a mile from the reef, in eighteen fathoms."

It was thought after this escape, considering also the valuable cargo our ship had on board, to be too hazardous to continue any farther in a route so entirely new and untraversed. We therefore hauled the ship up to the northward, and soon after getting into the usual track of the Spanish Manilla ships, steered a course for the Island of Tinian; consequently in a path so often sailed over, we were not in any probability likely to make any farther discoveries while on this passage.

June 16th. Had a strong gale or fresh trade from the north-east, accompanied at intervals with squalls. This day we made use of the last of our stock of plantains and bananas, laid in at Nuggoheeva; had yet, however, so good a store of cocoa-nuts, yams, and potatoes remaining, that every man received one each per day. Latitude at noon, 9° 35′ north, longitude 167° 14′ west. Among the other birds around the ship this day, we saw a Port Egmont or Cape Hen. Our route, from about five degrees to the south of the equator to our present situation, must be excellent sperm whale ground, for scarcely a day has passed, but we had sight of those valuable fish, oft-times in very great shoals, more particularly about the equator, and in the vicinity of Fanning's Island and the Washington group.

June 29th. Had pleasant weather, attended with a light trade wind: the ship was this morning surrounded with the man-of-war hawks, boobies, tropics, and a variety of small birds, among which was a species much like a small white gull; from this we were induced to believe that we had passed Gasper's Island sometime during the latter part of the past night. Latitude, at noon, 14° 48′ north, longitude 192° 23′ west. This day tried and found the current to set N. by E. at the rate of half a mile per hour: for the several days past it has been northerly.

July 4th. We had a light trade breeze, with delightful weather, all of which was very fortunate for us, as by this means we were, in some small degree at least, and to the best of our capacities, keen appetites, and plenty of edibles considered, enabled to add to the rejoicings at home our mite, in the good old fashioned way of enjoying a holiday. Butchered our last Nuggoheeva hog, and with a full allowance of fresh pork, yams, sweet potatoes, cocoa-nuts, plum pudding, and the like savory dishes, managed matters to have a pretty jovial time of it, topping all off with a moderate glass to prevent our choking. "Hail Columbia! happy land," concluding the feast, and bringing therewith very forcibly to our minds, the thoughts of "Home, sweet home!" In the evening, a noddy lit on the yawl boat in the tackles at the ship's stern, and suffered itself to be taken, apparently nothing loth to assist in celebrating this our national anniversary of the Declaration of Independence. After suitably feeding, we allowed him to have his liberty again.

On the fourteenth, had a strong trade wind from E. by S. Had had during the night, squalls of rain; at sunrise, however, it cleared up, and gave us our former fair weather. At 10 A. M. the seaman at the mast-head on the lookout, gave the welcome cry of "Land ho!" bearing W. by N. distant about seven leagues. This proved to be the Island of Tinian. At meridian, the south point of Tinian bore W. by N. half N. three leagues distant; made preparation then, by bending the cables, to bring the ship to an anchor. At 2 P. M. we were abreast of the south point, and began to open the bay at the southwest part of the island, where, laying at the inner side of the reef, near to the shore, a wreck was brought to our view; upon which discovery, we set the American colors. An hour after, brought the ship to anchor in seventeen fathoms of water, over a bottom of coral rock and sand; the south point of the island then bearing S. by E. and the north-west point N. W. half N. After having the ship's sails furled, a boat was hoisted out, manned and

VIEW OF THE ISLAND OF TINIAN, LADRONE ISLANDS

From an engraving in Anson's *Voyage Round the World*, London, 1748

A SHIP OF THE EAST INDIA COMPANY
From an etching by E. W. Cooke

armed, and proceeded for the shore, where several men could be seen near the wreck, which, as we drew towards it, proved to be that of a ship of between three and four hundred tons, bilged, and laying over on her side. While passing through the passage in the reef by the wreck, it had been noticed that many of the persons on the beach were Malays, and it was therefore thought to be at least the most prudent step for us to be on our guard, and make use of all proper caution in approaching them. As our boat came within a few yards of the shore, the men ceased rowing and lay on their oars, until it was ascertained who they were. On hailing the person who appeared to have command, motioned to his men to fall back, at the same time himself coming forward, and answering in English, that he was the commanding officer of the crew belonging to the wrecked vessel: he, advanced to meet us, as we landed from the boat, and after shaking hands, gave me to understand his name to be Swain, that he was an American, and was born in Nantucket, held the station of first officer on board the lost vessel, and, since the death of the captain, was of course in the chief command, soliciting my assistance in behalf of himself and shipwrecked companions; to this I could but reply, that as it was the bounden duty of every man to render all the assistance in his power, to his fellow-creatures in distress, he with his friends might rely upon receiving mine.

He then proceeded, in answer to my inquiries, how many persons there were in all of his company, to state, there were three females, and twenty-one males, viz., the captain's widow, her servant woman, and a female infant, two years of age; three officers, six British seamen, nine Lascars, and eleven Malays; while as he thus finished his statement, giving us an invitation to visit their habitations. These we found pleasantly situated on a beautiful lawn, surrounded by, and having at suitable distances about the same, stately trees, the whole appearing to be a cool and most delightful residence. The first house we arrived at was that belonging to the captain's widow, to whom I was now introduced by Mr. Swain: she was a very lady-like woman, of an easy and graceful demeanor, about thirty years of age, at the moment somewhat unwell, in consequence of the shock she had experienced, from our vessel's coming so suddenly to view, and although getting gradually more composed, yet considerable anxiety still remained depicted upon her countenance. From Mr. Swain it appeared, that this lady was engaged upon some household matters in doors, while the servant woman, at the time busy in front of the house, was the first to notice the approach of our ship by the south head of the bay; it was this woman's exclaiming, "A ship! A ship!" that brought the mistress to the door, who, on beholding a vessel so near, and under full sail towards them, swooned away, and fell to the floor, nor was she brought to herself again until we were on the point of anchoring; not neglecting, while thus recovering, to offer up a thankful prayer to Heaven, for so bright a prospect of deliverance from her present situation, and a restoration once more to her country and friends. The lady herself observed, that the moment our flag was seen flying at the mizen, it at once told them the stranger belonged to a Christian country; this of itself being a sufficient guaranty that her commander would not refuse herself and child a passage to her friends. I assured her, that together with the infant and servant maid, she was very welcome to, and should have as comfortable accommodations as our little ship could possibly afford, while on the passage. With many expressions of thankfulness on her part, for thus answering her expectations, we were invited to take a cup of tea previous to returning to the vessel, which of course we did not decline, but accepted with pleasure; and then proceeded to the next house, which was the officer's lodge: nearly in a line with these, were three others, for the seamen, Lascars, and Malays, the whole forming an oblong square, and erected by means of piles driven into the ground, other pieces connecting these at the tops, having a sharp roof thatched with palm leaves, and the turf flooring kept neat and clean by means of mats, of which they had great plenty, made them very comfortable; the men appeared to be under entire subjection to their officers, and quite content.

Within a few feet of the most inland lodge, was a well of the aborigines, or ancient inhabitants of the island; this, walled up in a very neat manner with hewn stone, tapering from the top to the bottom, was fifteen feet diameter at the top, and five at the bottom, with a flight of thirty-six stone steps on its south side, descending to the water, which was very good. Near by this, piled up in an oblong heap, under cover of the ship's sails, was her valuable cargo of silks, teas, &c., and buried in an appropriate grove, on the side of the lawn near to their habitations, were the remains of their captain.

The habitation of Mrs. McClannon (the captain's widow), where we now repaired to fulfill our engagement, was found in very neat order: it was about twenty-four feet by twelve, the inner walls being hung round with blue nankeen, a screen of the white, separating the farther end of the room into a lodging apartment. We found the tea table already set, and most bountifully furnished with what was a very agreeable picnic for persons from a long voyage, viz., baked bread, fruit, broiled chickens, beef steaks, and China sweetmeats, to which was added an excellent cup of tea; events during their residence on the island, and other agreeable conversation, soon bringing the hour for returning to the ship at hand.

From the following particulars, as given me, at my request, by Mr. Swain, it appeared, that their ship was an English vessel, the annual supply ship from the Honorable British East India Company at Canton, for the British settlement at Sidney in New South Wales, with a full cargo on board, consisting of teas, silks, nankeens, China ware, sugar, rice, sam shu (a Chinese liquor), ginger, candy, and spices. They had, on leaving Macao, crossed the China Sea, and passed Formosa, when on gaining the longitude of Japan, were met by severe storms, so straining to the ship as caused her to leak badly. When built, this ship had been iron fastened, then sheathed with inch boards, put on with iron nails, her bottom coppered over this sheathing; this, at the time they encountered the gales and storms spoken of, was worn thin, and was continually breaking and pealing off, while the iron nails by which the board sheathing was fastened, were eaten off by the copper, so that the sheathing would start and come from the ship's bottom, thus leaving the main plank exposed; the oakum had also, from long standing, become defected, and from the ship's motion, washed out of the seams, so that the leaks rapidly increased; to remedy this unfortunate state of things, their captain had judged it most prudent to bear away for this island, supposing that by laying her on the beach within the reef, she might be repaired.

Arrived at this island, they became fully satisfied, after a survey and inspection, that it was utterly impossible for them to prosecute their voyage any farther, until the seams were recalked, as then she was only kept from sinking by continual pumping. To accomplish their determination of warping the ship through, and within the reef, then to take out her cargo previous to hauling her on the beach, to proceed with the needed repairs, they had just commenced, as a gale arose, by which the ship was cast on the reef where she now lay, and bilged; not, however, until they had succeeded in getting out the major part of the cargo.

It was now upward of thirteen months since this ship had been cast away; the crew, however, were so happy as to be thrown on one of the most fertile spots in the world; they had tamed a milk cow, a few young cattle, a number of swine, and domestic fowls; these all run wild. They had, moreover, a large supply of bread-fruit, cocoa-nuts, and many other excellent tropical fruits, besides which, their cargo, from its assorted character, furnished them with suitable clothing, as well as some of the most to be wished for articles for the table; so that they were only to be considered as suffering from the want of society. This had been the case with the lady more particularly, since the misfortune and its consequent fatigue had so pressed upon his mind, as to throw the captain into a violent fever, which in a very few days had caused his death.

A hunting party of Malays and Lascars was sent out by Mr. Swain, who had the goodness to offer us some fresh beef, to procure from among the wild herd a young heifer for this purpose. Noticing that they took only a couple of sharp hangers and knives, I inquired what they were going to do with these: in reply, Mr. Swain related their manner of taking these animals. The prairies where the numerous herds of wild cattle are in the habit of grazing, are covered with rich feed; around these are thick woods, impenetrable by reason of the underbrush, reed, vines, &c., excepting by such paths as the cattle had worn in passing to and from one prairie to another; these paths are so narrow as to oblige the cattle to go along in single file. Two men were usually posted in ambush in these narrow ways, some few feet apart, so that should the first one miss giving a deadly stroke, the second would complete the business: these being stationed, the others of the party make a circuit at a suitable distance, until they were arrived on the opposite side of the prairie in sight of the herd, when driving them off, the cattle always following the leader (a large bull), in Indian file, went by the two first mentioned men;

they were allowed to pass until an animal to their liking was approaching, when the first man would strike at and endeavor to cut the ham strings; if he is so fortunate as to cut both, the creature instantly falls, and is butchered on the spot, the meat being cut into pieces sufficient for a load for each person; if not, the second seldom fails in accomplishing this end.

The party sent out this time, returned in less than two hours, with the several portions of a fine fat young animal, to be sent on board for the ship's company.

After returning to the ship, it became necessary to hold a consultation, to arrange sundry matters, and make some suitable preparations for the reception of our new and unfortunate friends, and their accommodations during the passage to Macao; this we were enabled to do by dividing our after cabin, for the greater comfort of the females, into two parts, the other being reserved for the officers; afterwards receiving six boxes and trunks of silk goods, which, together with their luggage, considering the size of our vessel and her full cargo, was all that could be taken on board, and these only for the benefit and by the desire of the widow and officers.

July 17th. The same pleasant weather which has favored us so long still continues, as well as the moderate trade wind. Having obtained wood, water, some refreshments, and an addition to our stock of provisions, of some rice, tea, sugar, &c. a young bullock, several hogs, and poultry, we were now well furnished for the passage to Macao.

It was upon Mr. Swain's proposition, decided to take the Lascars along, and leave the Malays on the island, in consequence of a quarrel that had taken place while the boats were employed in transporting our passengers on board ship with the seamen and their luggage, between them, in which the Malays, intoxicated with the *sam shu,* had fallen upon the Lascars, and beat and wounded one of the latter seriously, by stabbing him with a cruse or dagger, so that he became faint and feeble from loss of blood, before we were able to get him on board and dress his wounds; he, however, by good nursing and attendance gradually recovered, and by the time of our arrival at Macoa was nearly healed. This we the more readily acceded to, because that it was Mr. Swain's intention to charter a vessel, and return from Macao to this island for the cargo; the Malays, moreover, would have plenty of provisions, and even the comforts of life, so that there could be no likelihood of want overtaking them previous to his return, to which a farther consideration was the smallness of our ship, the quantity of water she could with other things carry, being only a moderate allowance for the company with a fair passage, exclusive of the Malays.

Being all ready for sea, we weighed anchor and made sail, leaving the eleven Malays in possession of the island, as well as in charge of the wreck and cargo, until Mr. Swain's return, which was in the space of about five months, as I subsequently learned. Until we drew near the entrance of the China Sea, the weather continued much the same as for many days back, at intervals with heavy squalls of wind and rain, though generally with a brisk trade wind.

August 3d. Our lookout gave the welcome notice

of land in the W. S. W. about ten leagues off; this proved to be two of the Babuyane Islands, the wind at the time a light breeze from the E. S. E., weather very pleasant. During the night the wind veered around to the S. S. E. and at noon the next day we were about midway between the two northward and eastward islands, the southernmost bearing S. S. E., the northernmost N. N. E. A great deal of drift stuff was now about the ship, such as trees, bamboo, straw, and grass weed. Our observed latitude at this time was 19° 28' north, having made 23° 41' west longitude from Tinian. At 1 P. M. had sight of Cape Baxadore, on Luconia, bearing S. by E. twelve leagues distant, at the same time the westernmost island of the Babuyane group bore E. by S. half S.

These islands are twelve in number, and are all elevated or high land; from the N. E. part of them a long reef of rocks puts off in the same direction; some of these are so prominent as at the distance of four leagues or thereabouts to appear very much like ships under sail. Among these islands we found strong irregular currents. The ship's course was now to cross the China Sea, for Macao.

August 5th. The winds were light and variable, the weather still pleasant, the sea as smooth as a mirror, here and there parcels of drift, trees, &c. with many dolphin and other fish about the ship. By the fifth day after we had made our passage across the China Sea, having then a moderate royal studding sail breeze from the eastward; this, at eight in the morning, brought us in sight of the grand Ladrone Island, then bearing N. half E. about ten leagues distant; on sounding, we had thirty-eight fathoms water, soft muddy bottom. At two P. M. the wind dying away, we came to anchor in twenty fathoms; at day-break, weighed anchor again, the wind from the E. by N. and stood in shore, where we received a Chinese pilot. The wind soon after veering around ahead, brought us once more to anchor in eight fathoms muddy bottom, the grand Ladrone bearing east, the island, with the representation of a ship's mizen on it, then bearing W. by S. There were a number of Chinese fishing vessels around our ship just at this time; many, as they sailed along, dragging extended between two of their boats a lengthy seine. One of these couple, during the middle watch of the night, was forced by the current athwart our hawse, so that the net or seine as extended, got foul of the cable, and in their anxiety to clear themselves without our knowledge, or giving us any alarm, they neglected to answer the hail of our Chinese pilot, whom the officer of the watch had requested to find out the object the two vessels had in view in taking that station; the poor pilot, receiving no answer, was sadly frightened, and instantly declared them to be Ladrones (pirates), who were cutting away the cable, to tow the ship ashore. This report the officer thought proper to give me, and as some addition to the confusion, the order to call the men to quarters greatly alarmed the females, and brought them in the cabin; yet for their terror, there was some cause, as previous to the ship's sailing from Macao on their late unfortunate voyage, the British East India Company's packet, bound from Manilla to Macao, had been captured in a situation near to our own, by a party of

MACAO, CHINA

From an oil painting by a Chinese artist, at the Peabody Museum, Salem

BOCA TIGRIS, OR THE "TIGER'S MOUTH," AT THE MOUTH OF THE RIVER LEADING TO CANTON, CHINA

From an oil painting by a Chinese artist showing the forts and the American ship "Telahoupa"

those Ladrones or pirates, and after a severely contested action, all on board, with the exception of two, were put to death: this affair coming fresh to the widow's mind, and hearing the above order, induced her to conceive a repetition of it, or at least a severe struggle was about to take place. When upon deck, I found the pilot greatly concerned about his head, which he was certain the Ladrones would cut off, and not answering his call was sufficient to make him believe these, were pirates. By getting a spring to the cable, to veer it, and bring our battery to bear, we were ready to settle the matter very shortly; but unwilling, through mistake, however sure the pilot might be, to injure any subject of the Chinese emperor, we hailed them again. Still no reply: the sentry was then ordered to discharge his piece over their boats, so that no one might be injured; on thus doing, the spell was soon broken; for their women began screaming and crying in great style, giving us to know that they were fishermen, whose nets were entangled with the ship's cable, which they were endeavoring to clear. We found, indeed, that they were only poor fishermen, who had on board their several families. At day break, when the cable was hove in, their net was found so entangled and wound around, that we were obliged to cut it into several pieces before it was cleared; this was a source of much grief to the two skippers (but it could not be avoided), for now they could not procure any *chow chow* (victuals) until the net was first taken on shore and mended. We explained to them the imprudence of their conduct, and the risk they had run in not answering, while at the same time, to assist the poor creatures, a *cum shaw* (gift) of rice, sugar, and some provisions was given them, and to their children a Spanish dollar each, as a fund to repair their damage: this produced in return many appeals to *Chin Chin ing Josh* (the name of their God) for blessings in our behalf, the women patting their little ones, and thanking us as long as their voices could be heard.

CHAPTER XIII

AT MACAO AND CANTON

EARLY in the morning of August 13th, 1798, our pilot had the ship under way again; wind light from S. S. E. and at 8 P. M. brought her to an anchor in Macao road, having the city in full view before us. After breakfast the following morning, accompanied by Mr. Swain, I landed at the city, where for a commencement I was met by an unexpected difficulty, and one that at first was like to have caused a vast deal of trouble before it was removed; this was because we had on board the English female passengers, whom the governor, a mandarin of high grade, declared he would not only not allow to land, but must also refuse a *chop* (permit) and pilot to enable us to proceed to Canton. Their China custom, which amounts to law, requiring the ship to depart and carry them away, *chop chop* (immediately). "It no have China custom; how can strange woman come on shore in Chinese country? No—can—do, loo—this act," conclusively argued the mandarin, as he turned upon his heel; the matter in his opinion admitting of no farther discussion. Not much enlightened, or greatly pleased with this sublime reasoning, I returned on board, rather heavy hearted at so dark a beginning, but nevertheless determined to try again. Next morning, waited upon the mandarin again, but found him as stubborn as ever, and for my farther information, was told that the decree of the Celestial Emperor, required all ships bringing foreign females, to give heavy bonds, that they would take them away again when they sailed, misfortune and distress, as in the present instance, being of no consequence; they could not be permitted to land, until these bonds were given, even at Macao, a Portuguese city, lest by some means they should get to Canton by land.

The Hon. Mr. Hall, President of the Council of the British East India Company's Factory at Canton, was then at Macao; by him I was received with much cordiality, and after expressing many thanks for my conduct, he kindly assisted me, and also made every possible exertion to obtain relief, and liberate our ship; still without the smallest probability of success. Nothing was sufficient to induce these officers to vary or make any allowance for a case (as this) not contemplated by their laws. Afraid of having his head taken off, the mandarin always replied, "It no have China custom; how can, do." Thus the second, and third, and fourth days in like manner were passed, with just about as much encouragement on the last, as on the first, and moreover, being daily harassed by the prevarications of the stubborn and unfeeling mandarins, governed as they were by illiberal laws and customs, made my situation exceedingly unpleasant.

On the fifth day, the case was finally arranged by Mr. Hall, who made the mandarin a handsome *cum shaw* (present), and giving bonds that the first English vessel or Company's ship that sailed, should take the females away. After this, a *chop* was issued for the landing of the females, as well as the officers and men, from the wrecked vessel, together with their luggage and cases of silk, and another for our ship to receive a pilot and proceed to Canton. While employed clearing ship and preparing to get under way, Mr. Parry, one of the supercargoes, came alongside in the President's yacht, a most superb schooner of about eighty to one hundred tons burthen, with a request from him, for a passage to Canton in our vessel, for a Mr. McKenzie, brother to one of their Council, who wished to get up to Canton, where he might have the attendance of his brother and the chief surgeon, having but recently arrived from Batavia, where he had been confined by a fever, though now so far convalescent, as to be able to walk about, with a little assistance. Mr. Parry farther stated, that there could be no danger of Mr. McKenzie's communicating the fever, as he had been many days with them, and no symptoms of others having received it could be discovered, as indeed had there been the least risk, the application would not have been made; it had, however, been refused by the captains of two of their ships, and by the captain of one from Philadelphia. I could not but answer then, that the case seemed somewhat hazardous, by the refusal of these captains, yet, although our cabin was small, the gentleman should be accommodated with a passage, and made as comfortable as our means would admit of. After expressing his acknowledgments in the name of the President, Mr. Parry returned to Macao, to bring Mr. McKenzie on board our ship; this he shortly did, having in company an Armenian, and Persian gentleman, likewise friends of Mr. Hall, who requested a like favor in their behalf; this was also assented to; Mr. McKenzie I soon found with the loan of an arm, was able to walk about with comparative ease. Mr. Parry* now took leave of us, and we having a Chinese pilot, and *chop* on board, hove up our anchor, and steered for Lintin, to pass the Boca or mouth of the river Tigris, for Canton.

August 21st. At 7 P. M. we came to an anchor in the bay opposite a fort on the east side of the Boca; wind moderate and variable, with pleasant weather. At nine next morning, while waiting for the mandarin to come on board and examine the ship, before giving our pilot a new *chop* to pass up the river, accompanied by Mr. McKenzie and the other two gentlemen, I took advantage of the clear and pleasant weather and landed near the fort, where an under officer of the garrison met us, who, had it not have been for our Armenian passenger possessing a smattering knowledge of the Chinese language, would have stayed farther proceedings in our intended walk; this gentleman, however, was able to negociate a treaty, and despatched the officer to the mandarin in command at the fort, who, in consideration of the small sum, or *cum shaw* of a Spanish dollar, not only gave permission to take our walk, but also directed the officer to show us to the fort, where on entering, his mandarinship, with his hands closely clenched together in front, and thus moving them quickly up and down, with the body

*Of this gentleman I cannot speak but in terms of highest admiration and respect; his exertions to obtain the release of our ship from the Chinese authorities, and in behalf of our distressed passengers, were not omitted by day or by night. An acquaintance with him then commenced, subsequently increased, will ever be held in remembrance by the author.

slightly inclining forward, actions on the whole somewhat resembling the shaking of a person afflicted with the paralysis, welcomed us with a *chin ching* (How d' ye do?) making the while a great many bows, in rapid succession, and as many professions of friendship.

We were then treated in the usual custom of Chinese politeness, with tea and sweetmeats, and afterwards allowed to look round the fort, attended by the officer who had first met us. There were in it fourteen handsome brass nine pound cannon, but all very uncouthly mounted: it was besides difficult to depress or elevate these pieces many degrees: officers and men were similarly accoutred, except a ball in the officers caps. Their military discipline, so far as we were able to judge by the specimens shown, was very far from being the best in the world. After leave taking was over, with this commander, who, according to their notion, was very polite, and to ours very friendly, we strolled as far as the top of a hill near by, from which a considerable view of the country was had; this appeared, especially the rich and extensive padda grounds in the valleys, to be pouring forth its productions in great abundance, and promising an ample harvest to its owners.

After we had left the shore a few rods, Mr. McKenzie seeing a fisherman carrying some good sized fish of the mullet kind, from his boat up to the fort, expressed a wish to have some, observing, he thought with one of these barbacued, he could make a most hearty dinner. To gratify his desire, I answered, we would immediately return and endeavor to procure one, but this he informed me would be fruitless, for it could not be effected, save through the agency of one of their compredores. We had scarcely taken a dozen strokes on our way to the ship, when a mullet, weighing fourteen pounds, sprang from the water to the height of our heads, and fell into the boat in the midst of us, as we were seated in the stern sheets, and while to prevent his flouncing disturbing our invalid, I had placed my feet on the fish, I was a little surprised to hear Mr. McKenzie, astonished to have his wish thus wonderfully gratified, observe, "Mr. Fanning, you certainly are one of Heaven's favorites, for no sooner is the wish of your friends made known, than an invisible hand places the means in your power to gratify the same." Be that as it may, the prize certainly came very opportunely. This fish, the steward cooked for dinner, agreeable to Mr. McKenzie's wish, and on it he made a very hearty meal, without (considering his health), experiencing any ill effects from the indulgence.

At half past eleven A. M. the mandarin had given the pilot a *chop* to go up. We accordingly weighed anchor, and with a fair breeze passed the Boca. The 23d, at two in the afternoon, we came to anchor at Wampoa, ten miles below the city of Canton, the usual anchoring ground assigned by this government to all foreign vessels.

The Commodore of the British East India Company's fleet, who was here in a fifty gun ship, had received a letter from the President of the Company, by a Chinese passage boat advising him of Mr. McKenzie's being on board our ship, and before our sails were furled, he had sent a lieutenant in his barge, to take Mr. McKenzie and the other two passengers, with an invitation, couched in the most polite and friendly terms, for myself to accompany these gentlemen, to his ship. This I accepted, and was received in a like flattering manner, being at the same time introduced to the captains of several of their ships, and his own officers, the commodore desiring me to consider myself as much at home on board his ship at all times, as in my own, at the same time expressing himself perfectly willing to render any friendly assistance in his power, that I might stand in need of; while, in case of his absence, the same would be promptly attended to by the then commander of their fleet. I returned him many thanks for his politeness and courtesy, and shortly after returned in his barge on board the *Betsey*.

At eight the next morning, having procured another *chop* from the mandarin, in the Chinese guardboat made fast at our quarter, with the pilot as a guide, I left the vessel in my own boat, for the city of Canton, where, after a series of tedious and vexatious examinations at five chop houses on the way up, I arrived in three hours time. Mr. Hall in a few days came up from Macao, from whom, together with the members of his council and several supercargoes, I received much assistance in my business; their attentions proving to be of vast benefit to my principals and to myself, being entirely unacquainted with the mode in which trade was carried on; and having received a kind invitation from the President to make myself at home at their factory, nor to hesitate at all to call upon him, or any gentleman connected therewith, for advice, I made, therefore, almost every day, a visit to this establishment, spending many an agreeable hour with my friend, Mr. McKenzie, who had an apartment to himself, and whom I found to be an amiable and intelligent traveller.

The usages and customs of trade at Canton, make it easy for supercargoes to attend to their business there with despatch; in fact, more so than at any port of the world I have visited. The first thing to be done, is to hire a factory, and thither the Chinese merchants and traders will all assemble, bringing with them the samples of what they have to dispose. This factory contains an audience or dining hall, lodging and store rooms, together with accommodations for the compredore (steward), servants, cooks, and coolies (laborers). After the factory is obtained, a compredore is engaged, then a trusty servant, who speaks the stranger's language, and attends upon your person in your walks, to act as interpreter. After this the ship must be secured with one of the Chinese hong merchants (i. e. upon receiving security, he agrees to pay all the duties, charges, &c.), of whom there are twelve, being an office answering to that of our collector. He grants all the *chops* (permits) for the cargo to be brought up to town, and also for the return cargo to be taken on board. This merchant will frequently, when making such an agreement, buy the bulk of the cargo, giving at prices then fixed upon, such portions of a return lading as may suit.

One afternoon, when taking a walk with the inter-

THE FACTORIES AT CANTON, CHINA
From an oil painting by a Chinese artist, at the Peabody Museum, Salem

MODEL OF A MALAY PIRATICAL PROA FROM THE EASTERN COAST
OF SUMATRA
From a photograph of the original (before 1838) at the Peabody Museum, Salem

preting servant, to one of their houses of worship, which was situated in the farther parts of the suburbs, near to the city wall, we came to the high wall inclosing the grounds around it. From this wall to the house, there was an arched entry-way or covered passage, twenty feet in width, and about twenty-five in height, having at both ends a gate, or rather large door, thirty feet apart (the distance from the street to the building). Under the arch, and directly over the inner door, was placed the gilt figure of a very corpulent man, having on each side of it a *non descript*, the upper part of whose bodies resembled those of very large Africans, grinning most hideously, thereby showing their red gums and white teeth, while from the waist downward, they had something of the shape of an alligator. Who does this great figure represent? I inquired of the Chinese servant, as we stopped, looking a while at them. "That have Josh," (God) he answered, with emphasis. Well, if that have Josh, who are the two on each side meant for? "Them other two," continued he, still more emphatically, "on both sides of Josh, have mean for the devil." What have they put devils so near Josh for? "Oh!" said he, "that be for to take care and guard Josh, and see no man hurt him."

The house is on the back part of the grounds, and is supported in front by pillars, placed some few feet apart; between these, are folding doors of lattice work, which in time of service are turned back against the pillars, so that the worshipping people, as collected outside the iron railing (running round the front and sides of the house), have a fair view of the services, as they are being performed by the priests.

The interior of the building consists of one spacious room, where, near the centre of the back wall, and against the same, considerably elevated, is the image of Josh (their God; no devils being necessary for his defense where the priests are) a very fat and portly personage, in size equal to four common men, and most splendidly gilded. Immediately in front, and raised to a level with its feet, is the altar, on which Josh wood (sandal wood) is kept burning day and night: by this, the house and about, is filled with a most delightful fragrance, esteemed by the people as sacred. A few feet from the altar, and about the centre of the room, stands a capacious table: on this are deposited the offerings of the inhabitants, sent in during the day, to and for the support of Josh; these consist of the choicest of their cooked meats, pastry, fruit, viands, &c.: thus, if a person has a pig or the like for dinner, the half is sent to this place, the servant returning for the plate after Josh has eaten. A little off from the altar is a small door, scarcely large enough for a man to get through, opening to a secret passage, which leads to the dwellings of the priests; these are erected against the walls, where all the etables are conveyed during the night, and appropriated for their especial benefit. In this way they are enabled to live well; they have also sufficient address to make the people (no difficult task, however, because of the latter's stupidity), believe that Josh has devoured all and is much pleased therewith, and that the *chin chin* (sacrifice) has been a good one.

On commencing their devotions, the priests, to the number of fifty (as each Josh house has more or less), form in single file in front of Josh, each one by his mat, on which they kneel, and bow down to the image, and kiss the dust; after this, all rise and follow their leader round the room, marching in a circle between the altar and table, all the while chanting; sometimes they go nine, others but three times round; then kneel again on their mats, and kiss the dust as often as they make the march about the room.

There were forty priests attached to this house, from twelve years old to seventy and upwards. I was told by the head priest, that every one of their order, belonging to all the Josh houses in the empire, received an annual pension from the emperor for their support; the young priests were then learning, and only when grown to manhood, could they *chin chin* to Josh: both young and old, are all prohibited from marrying. A foreigner, by a present of one or two pieces of silver to the head priest, can obtain every information concerning their mode of worship, living, &c., which he desires.

Having disposed of our cargo, and by the 23d of October, obtained in return another, of teas, silks, nankeens, China ware, &c., we were soon in readiness for sea; but upon the proposition of a captain bound for Philadelphia, agreed to wait until his vessel was ready, it being thought by keeping in company, until we should pass Java Head, would be for our additional security against the attacks of Ladrone and Malay pirates, who were numerous at this time. This ship also carried heavier metal than my own, and I therefore thought it prudent to stop and sail with her. The *Ontario,* Captain J. Whetten, arrived at Canton while we were there, according to expectation, and was left by us.

CHAPTER XIV

CANTON TO NEW YORK

OCTOBER 30th, 1798. In company with our consort-ship, at 3 P. M. we passed the city of Macao, and steered to sea (our consort bound to Philadelphia), under a moderate breeze from E. by N.

December 6th. Our consort in company, passed the Island of Thought-the-way, with the wind from the southward. Nothing unusual had occurred, except that for some days our vessels had been watched by several piratical proas, who were continually dodging us and keeping aloof; we however, kept on our way through the strait of Sunda. On passing the south point of Sumatra, our consort then half a mile to windward, we opened a bay on the coast, by doing which, a fleet of piratical proas were discovered drawn up to meet us: part of these we had before seen, though the number was now increased to twenty-nine. These fellows putting on a bold and defying front, judging from the manner in which their fleet lay posted behind this point, made it soon become evident that they were waiting an attack from the ship.

To our surprise, the Philadelphia ship, so soon as she had the pirates in view, hauled on a wind for the Java shore, and being a far superior sailer to the *Betsey* on a wind, soon left her, widening the distance between us very rapidly, disregarding our signals for them to stand by, as well as the agreement for mutual protection entered into before starting from Canton. This did not pass unobserved by the pirates, but emboldened by this separation, and confident in the capture of (ours) the smaller ship, they quit their anchorage and gave chase after us, in three divisions, making use of all their sails and oars, and were very evidently gaining upon us, never ceasing as they came on, to pass signals of some kind from one division to another.

Unfortunately, at this moment the wind began to abate, and finally failed us altogether, so that our ship, upon a perfectly smooth sea, moved only at the rate of one and a half miles per hour. The little preparation we could, was made to give them a warm and hearty reception; all hands were at quarters, and with our eight four-pounders of iron, and two brass long six-pound guns, each charged with a round shot and bag of musket balls, we waited the approach of these marauders.

As a word of encouragement, I stated to the officers and men, that inasmuch as our consort had made good his escape, there was now no resource left but to defend the ship to the last, yet if every man was firm and undaunted, obeying orders and doing his duty as a freeman, there was at least a glimmering of hope that we should come off with flying colors; but should there, on the contrary, be any flinching, death by the cimetar or poisoned cruse, as usually dealt out by these villains, was certainly in store for us.

The enemy continued his pursuit (as we still kept on our wind) by pushing forward one of their divisions directly under our stern, while within a quarter of a mile on each side of this were the other two divisions, seemingly advancing with the intention of falling on our bows, at the same time the centre should make the attack at the stern. It was to this last my attention was particularly directed; although judging from the signals, their leader or admiral appeared to be in the right wing, still this was the division likely to come first into action, and could it be by any means disabled before the others came on, we had some expectation the other two would hesitate, and thus give us a better chance for future operations, otherwise it was clearly evident a general onset from the three would overwhelm us with their immense superiority in numbers if no other way; to beat them in detail, therefore, was our plan.

On a nearer approach, they commenced shouting most tumultuously, and opened their fire upon us; the centre division by this time was within musket shot distance, and discovered a set of some of the most hideous animals that ever the light of the sun shone upon; to add to this savage appearance, as well as with a view of intimidating our crew, they increased their yellings, to a rate that would have been creditable to the lungs of a war party of wild Indians. At this moment I clapped the helm a weather, hauled up the courses, and the ship, quickly wearing off, brought her broadside as handsomely as mortal could wish, to bear directly on the proas. We let them have it, in this the first discharge dismasting the centre vessel, and disabling two on each side of her; the effect produced was as expected; they instantly stopped their headway by means of their sweeps, and were apparently making up their minds as to, how next, we now wore ship again, and the better to assist their meditations, gave them another broadside with a suitable proportion of musketry.

Their admiral then concluded to make the signal for a retreat, which was very promptly obeyed by the whole body moving off with the disabled proas, leaving the dismasted one to our further good will and pleasure. By wearing ship first on one tack then on the other, we brought the broadsides alternately to bear, and delivered their contents in succession, the two brass pieces throwing with more force and farther, were shifted every time the ship wore, and directed upon the wings; for these pieces, while at Canton, some leaden balls had been cast, which I now found were thrown a third farther than the iron; we therefore kept them playing on the enemy so long as one could be seen to reach, for by this time their whole fleet were clearing out as fast as they could.

From our frequent wearings, we had finally got alongside the dismasted proa, and now grappled and hauled her alongside. Her crew had all quit and gone below, but when the boarding officer and party gained their deck, its commander came up, and kneeling, laid his head down, at the same time placing the officer's foot on his neck, in token of submission and with so very supplicating a look, that to relieve his fears, the latter rested the point of his hanger on the deck, and taking this submissive enemy by the hand, raised him up, then drawing the cruse from its scabbard at the pirate's waist, gave him to understand that all his men must come up and deliver their arms also, to him; this he readily understood, and at his call they severally appeared, one at a time, each delivering his arms as he came on deck. After taking and destroying all their armament, to insure which a strict search was made for the same throughout their vessel, we let them depart; an unexpected favor, and one which they acknowledged with many signs of thankfulness.

By the time we had again made sail on our course, the piratical fleet were quite out of sight, having entered a river or creek up the bay, and near to a town on the Sumatra coast, while far away in the southwest quarter, not to be seen from the deck, and only faintly discernable from the mast-head, was our valiant consort. I was sure that Mr. G. the supercargo of that ship (than whom no gentleman has a nicer sense of honor) could have had no hand in her leaving us at the time she did; for on all occasions, while at Canton, and since, he has proved himself to be at once a generous and real friend; it is therefore but justice that no blame should be attached to him in all this affair, as I have subsequently learned that he was very anxious to have their vessel bear up to our assistance, but could by no means prevail upon their captain so to do, his agreement, for it was his own proposition, remaining unheeded.

The next day we came to anchor at the harbor of the island of Cracatoa, and on the succeeding day a small Dutch despatch schooner came in from their settlement at Anjier, having on board as a passenger, an officer of the garrison at that place: of him we learned that a party of Malay traders from Sumatra, had been at Anjier, and made known the fact that a fleet of

Ladrone proas had attacked a vessel of war, their admiral having committed a great error in mistaking her for a small merchantman, by whom they had been defeated, losing many men and one proa; for this deficiency in judgment, their admiral had been broke, and another appointed in his place, they stating as a reason by which they knew this had been a man-of-war, was her sending her shot so far, which no trading ship could have done. It was this report of the Sumatra traders, and the probability that the ship would carry her prize and prisoners to Cracatoa, there to water, that had brought this officer; under the impression we would have the proa's crew as prisoners, he had come prepared to purchase them for slaves, and was willing to have paid on an average at the rate of three hundred dollars per head: he appeared to be sadly disappointed when informed that their liberty had been given to them again, "They are a bad race of fellows," said he, "and are far from deserving such liberal treatment."

Our ship, it will be remembered, showed fourteen guns; four, however, false, but so painted as exactly to resemble our iron ones. Her quarters were of man height, with hammoc and boarding netting fore and aft; she was painted all black, except a narrow red streak around her, and red ports, and withal, had a crew of twenty-seven men, who took some pride in giving their ship as warlike and man-of-war appearance as possible: this induced the Dutch officer to remark that he certainly would have taken our low ship for an armed cruizer. Previous to his returning to

Anjier, we purchased some green turtle, fruit, and fowls of him.

December 10th. At day-break weighed anchor; weather cloudy, with a fresh breeze from N. N. E. At 6 P. M. Java Head bore S. E. eight leagues distant, and from hence took our departure for New York. Our passage across the Indian Ocean, was unattended by anything more than the usual occurrences of similar voyages; watching the wind, trimming sails, making and mending, constituting our daily business.

January 6th, 1799. This sameness was rather uncomfortably and unprofitably relieved by the carelessness of a seaman, who had been set to watch the boiling of a small pot of pitch in the caboose, which the carpenter, who that day was busy on the yawl boat, had need for, to pay her seems with. The man not attending to his business, let this boil over and take fire, and with a view to carry it to the lee waist, caught the pot off, but in so doing burned his hand and let the whole fall upon the larboard deck; in an instant the whole, extending from abaft the mainmast to abreast the foremast, was in a bright flame. I was then seated in the cabin, but hearing the cry of "the ship is on fire!" and the man's screams, sprang to the deck, and had his hands bound up in a woolen jacket, while other blankets and woolen jackets passed up from below, were wet and spread over the flames, and around the main and fore rigging, and being kept wet, prevented the fire from running aloft, and finally extinguished it, not however until it had charred our deck, and burned through the side of the boat, stowed in the choks* and

*Chocks, i.e. wedges.

amidships. This was an unfortunate occurrence, and, by the force with which the flames raged, placed us for a time in a very perilous situation, distant as we were one thousand miles from any land. It had such an effect on my mind as to deter me ever since from suffering tar, pitch, rosin, or the like, to be heated on shipboard, at sea; I can earnestly recommend the same prohibition to sea captains, lest they should fall into a similar painful situation.

January 30th. Doubled the Cape of Good Hope, sounded, and got bottom on bank Lagullus, at one hundred and twenty fathoms, fine sand and shells. Saw many seals, some albatrosses and other oceanic birds, and far off in the horizon, to the windward on our larboard beam, a ship, which we took to be the same that had left us with the pirates.

February 3d, in latitude 33° 32' south, longitude 15° 10' east, were overtaken by a violent gale from N. W. by W. and obliged to lay to under storm sails for several hours.

February 24th. Passed the Island of St. Helena, under a fine trade wind from the S. E., and on the eighth day of March crossed the equator for the fourth time during the voyage. Arrived at the inner verge of the gulf stream, we were again overtaken by a violent storm from the N. E., and although under bare poles, our little ship lay over on such a rank heel, as not only to oblige us to send down on deck the topgallantmasts and all the light spars and booms from aloft, but to lower down the lower yards on the gun rail, and to saw away several stanchions and feet of the lee bulwarks, to free her from the weight of water on deck, before she would right, and lay free during the gale. After laying to for sixteen hours, the gale veering to the E. N. E. we bore up, and scud before it. The next day it so far moderated, as to enable us to sway up the fore-yard, and set a reefed fore-course; that following day the wind was round to the S. E. with heavy rain, yet more moderate; embraced the opportunity, and sent aloft our yards and masts again, and made sail on the ship.

April 18th. Had the wind fresh from the S. S. W. At two P. M. were on soundings; half past eleven A. M. our lookout gave the welcome cry of "Land ho!" This proved to be Long Island, eight or ten leagues to the eastward of Sandy Hook.

After an absence of two years from home, on a voyage around the world, or elsewhere, the feelings on obtaining sight of one's native land again, from which they have not heard during such time, are not to be expressed; thoughts upon a hundred different subjects fly also through the mind—a multitude of questions also arise, tending to give pain by the incapacity there exists satisfactorily to answer them—while with the utmost anxiety, the mind flies from one subject to another; these giving birth to others, are rapidly followed by those long dormant, but not forgotten remembrances of the condition in which all were left, hopes, doubts, fears and expectations rapidly succeeded each other. How will be found our near and dear relatives and friends? Who still remain among the living? How many, and which of them have gone on the long voyage of eternity? Who have been consigned to the tomb? and what shall we find the situation of our beloved country to be? All these questions and others, crowd through the mind at once, and remain unappeased until we once more gain the family circle, where they can be answered.

The wind had now shifted, and commenced blowing a strong gale from the W. N. W. It was thought prudent, therefore, to bear up for Montauk Point, and pass round the easternmost end of Long Island into the Sound; this the more especially, in consequence of three of our crew being confined in a very helpless condition by that dreadful disease the sea scurvy, to their hammocs; all on board, in fact, were more or less afflicted by it.

At midnight, had sight of the light on this Point; soon after passed by it, and hauled on a wind for the north shore. At nine o'clock next morning, the wind dying away, with the tide ahead, came to anchor abreast of Stonington, back of Watch Hill reef, in eighteen fathoms, Watch Hill Point one league distant, bearing N. E. by N. Sent the first officer in a boat to Stonington for a pilot, at the same time to procure some refreshments for our sick; in a few hours he returned with both these.

So soon as the tide became favorable, weighed anchor, and passing through Long Island Sound, arrived at the port of New York on the twenty-sixth day of April, 1799, after a passage of one hundred and seventy-eight days from Canton. This lengthy passage was owing principally to the fact of our ship not being coppered; her headway having been greatly impeded, since her departure from Java Head, by a foul bottom, the shell fish, marine grass, &c. adhering thereto.

While discharging our China cargo, some of the teas were found to be a little damaged by the gale we had experienced in the Gulf. Notwithstanding this drawback, when the same was sold, cost of ship and outfits deducted, as well as interest, insurance, and all the charges on the closing of the accounts, there was still remaining, to be divided among the owners, a net profit of $52,300. The amount paid into the national treasury as duties on our China cargo, was more than three times the cost of the ship and her outfits.

Thus successfully terminated the author's first voyage around the world, performed under the blessings of a kind superintending Providence, without the loss of a man; and this he believes to be the first American vessel, officered and manned wholly by native born citizens, that ever sailed round the world from the port of New York.

CHAPTER XV

AROUND THE WORLD IN THE "ASPASIA"

IN January of 1800, the *Aspasia,* a corvette-built ship, pierced for twenty-two guns, and just from the stocks, was purchased, and provisioned, and armed, by a company of gentlemen in New York, for an exploring and sealing expedition to the South Seas. Whatever, on so long and hazardous a voyage was thought would in any wise be connected with the health and comfort of the officers and crew, was furnished in the most abundant manner. Her equipment consisted of twenty-two very neat and handsome six and nine pound guns, with a suitable proportion of muskets, small arms, and all the munitions of war requisite for a ship of her size. She was commissioned by the government or president of the United States of America, as a letter of marque during the general war then raging among the nations, five officers holding the rank of lieutenant, a master, surgeon, eight midshipman, with a competent number of petty officers and men, composed the company, the whole under command of the author. Early in May, the ship hauled off in the stream, and thence dropped down to the quarantine ground at Staten Island, preparatory to starting.

May 11th, 1800. At 8 A. M. got the ship under weigh, and bore up to pass the Narrows, having a moderate breeze from the N. N. W. and pleasant weather;

this, as we passed the east bank, fell away to a calm, and was succeeded by a sea breeze from the southward, with which the ship was beat out to sea, accompanied by the British frigate *Cleopatra*, who, though at anchor as we crossed her stern, immediately hove to and filled away, keeping close under our lee, and within speaking distance. At 4 P. M. backed the topsails, and discharged our pilot, at the same time taking leave of a number of friends and acquaintances who had thus far accompanied us, and were now to return to the city in the pilot boat, after which trimmed the sails on the starboard tack, and stood to the eastward, our neighbor, the frigate, doing the same.

From this frigate several impressed American seamen had been taken by the authorities while she was at anchor under the guns of the fort on Governor's Island; doing which had very greatly displeased her commander, who, it was said, had threatened to retaliate, and replace these lost men from any American vessel, upon the first opportunity. After this transaction, a party of officers from each ship had unfortunately met at a public house on Staten Island, while the vessels lay at anchor near each other off the same place, and between whom many unfriendly observations had been exchanged, thus very greatly increasing their irritation; together with these was the fact, after we had beat over the bar, of our piping to quarters, spunging out the guns, and so securing the tompions as to keep the guns dry while at sea. This, though merely a precautionary measure, and performed without the slightest idea of offending, had, as I was afterwards informed by the pilot on board the frigate, been supposed to be, by her commander, our loading the pieces; and so doing, immediately under the muzzles of his own, was regarded by him as a presumptuous insult. It was the remembrance of the above disagreements, and seeing the present manœuvering of the frigate, that led me to suppose her commander had thought of replacing his men from our number. However, after duly considering the question, we came to the determination to stand upon the alert, and be prepared for the worst: should his boat be sent to examine the men (their usual manner of proceeding with impressments), the officer from the same was to be allowed to pass the gangway unaccompanied by any other person, and then made acquainted with this resolution, while with the colors flying we could act on the defensive, and keep all the force his boats should be able to bring against us, at a respectable distance, or if his guns should open upon us, a shot of ours in return, possibly might disable some of his spars; at all events, when the worst did come to the worst, we could "give it up," and let those in authority at home settle the matter. With these reflections on our part, we soon lost sight of the Highlands of Neversink, the *Cleopatra* off our larboard beam, at short musket-shot distance; in this way we proceeded under full sail, the two vessels sailing so nearly alike that in a pleasant southerly breeze, the bells of one ship could be heard on board the other.

The second day after, when nearly in the longitude of Halifax, for which place it was said the frigate was bound, she immediately, after a meridian altitude, and our ascertaining the latitude, bore up before the

THE BRITISH FRIGATE "CLEOPATRA," 32 GUNS

From an engraving in the *Naval Chronicle* (1805), after a drawing by Pocock

SHIP "ASPASIA," CAPE HORN BEARING NORTH BY EAST

From a lithograph in Fanning's *Voyages*, New York, 1833

wind, set studding sails, and very unceremoniously left us, without a parting compliment by word or signal.

While continuing our voyage, after crossing the equator, we fell in with the coast of Brazil, five or six leagues to the northward of Pernambuco; then beat up along the coast, and came to an anchor in the roads adjacent to that city. Shortly after this, accompanied by the surgeon and an officer, I proceeded in the barge to the town, entering through the narrow passage to the port, between the break-water, a work which must have been constructed at considerable expense, and the main shore. By the officer who received us when landing at a pier in the basin, we were conducted to the commandant at the fort, and thence to his Excellency the Governor. The reception was very polite and courteous: to his several inquiries about the character of our vessel, and the cause which had induced us to stop at their city; we gave answer, by stating, the *Aspasia* to be a private armed ship, commissioned by the government of the United States of America, and as bound to the South Seas and China, was in some want for sundry articles, required by her company on the lengthy voyage, which it was necessary to obtain before we proceeded farther. The governor expressed himself much pleased with the visit, it being by the first vessel of war of our nation ever at their port, and not only granted us liberty to obtain the supply of refreshments which had been our principal object in stopping, but at the same time directed an officer to show us the city, market, &c. After a hasty walk around this, we returned on board, but had not long been there, before a launch came alongside, bringing a present from his Excellency; this was the four quarters of a fat bullock, some excellent mutton, vegetables, fruit, &c.

A day or two after, myself and officers received an invitation from an aged nobleman, whose palace was situated a short distance beyond the bridge over the river, which passes through the southern part of the city, to spend an afternoon and evening with him; this we accepted. Our party was received at the front gate by his steward (a very polite personage), who led the way through the garden, where some little time was passed in viewing and inspecting his extensive menagerie of wild beasts, as also a numerous collection of the feathered race, from the Brazilian forests. Among others of very rich plumage, was the bird of Paradise. In the centre path of the garden, stood a fountain, from which several figures spouted the water in different directions, while immediately around, were trees bearing all kinds of tropical fruits, and a variety of shrubbery and flowers, filling the air with a most delightful fragrance; it was a most lovely spot, though but very poorly described.

We were then conducted to the palace, and introduced to its proprietor (whom the people of the city call the King of Pernambuco), an aged and very corpulent gentleman, moving about very heavily; he received us with much cordiality, and appeared to be pleased at our acceptance of his invitation, made a vast many inquiries respecting the institutions, laws, and government of our country, which our surgeon, who spoke Latin, was able satisfactorily to answer.

Cake, wine, a variety of dried and other fruits, were then handed round, and the evening passed off, with much gratification experienced by us at so agreeable an entertainment, as was spread before us by this sociable and hospitable nobleman. Nor was he the only person who manifested so friendly and courteous a bearing towards us; the same treatment was experienced at the hands of his Excellency the Governor, and the officers of the city, port, and garrison. Returning to the ship, after having received a supply of pigs, poultry, vegetables, fruit, &c., we weighed anchor and stood out to sea, highly gratified with our stop at Pernambuco.

While proceeding to the southward, we endeavored to get sight of the Island of Saxenburgh, but after passing over its situation as placed on the charts, and spending three days in search for it, without discovering the least sign of land, though favored by clear weather, we stood again to the south, perfectly satisfied that no island is in existence near the spot where Isle Saxenburgh is laid down on the charts. Our search might have been prolonged, but on our way we intended touching at Tristian de Cunha, to look for seals, as also at the Island of South Georgia, notwithstanding it was now mid-winter there. At this last place we expected to find the crew of a vessel called the *Regulator,* belonging to the owners of the *Aspasia,* which had been there cast away some months previous, and who might now be suffering from want of relief.

A few days after, in the evening had a distant view of Tristian de Cunha, the wind then blowing a strong gale direct from their bearing. In the midst of that dark night, the ship under double-reefed topsails and reefed courses, being a heavy press of sail, with a view to gain up to the land, or at least to hold our station, Mr. Sheffield, one of the midshipman, an active officer, and much beloved by his brother officers, in a heavy pitch, and lee lurch of the ship, fell from the lee poop deck overboard; the alarm was instantly given throughout the ship, and the main and mizen yards quickly thrown aback to the masts. This manœuvre in so heavy a gale, attended as it was by a cross-breaking sea, set her trembling as if the masts would come by the board; her headway, however, was immediately stopped by it; coops, spars, &c., were thrown over for our unfortunate shipmate, and the boat, as quickly as possible, lowered from the stern. In this was an officer and crew; but scarcely was she loosed from the tackles, before a heavy sea broke over, filled, and swamped her: this was a sad occurrence, and in a night dark as this was, little hope could be entertained of their rescue; fortunately at this moment, the ship under a press of back sail, had stern way on her; ropes were thrown over, and these all succeeded in catching hold of, and were thus hauled on deck, except poor Sheffield. For some moments, his cries away a stern could be heard above the noise of the elements; but soon all was silent, and though we kept the ship as near the spot as possible during the remainder of the night, under the faint hope of finding his body, at least, at daylight; still when this again shone around, not a vestige of him, of the boat, or anything thrown over, could be seen. This unhappy accident cast a gloom over the ship's company for a length of time.

At 8 A. M. all the three islands of Tristian de Cunha were in sight, when the wind moderating and becoming more favorable, we were enabled to work up towards them. At eleven, were close to Inaccessible Island, and sent an officer in the boat to examine it, who shortly after returned, finding landing altogether impracticable; the wind being now still more favorable, we stood for the largest or main island, and made it a league to the westward of the N. E. point on its north coast, thence steering alongshore to the eastward, we rounded the N. E. point, and discovered a small bay, where, as we landed on a beach at its head, lay a flock or rookery of sea-elephants. After an examination of the bay and shore in its vicinity, we proceeded some miles in the boat along the south shore, without discovering any seals, and then returned on board, making sail to the southwest. From our observations, it appeared that a ship might anchor and ride snugly at this little bay, with the wind to the westward, and at the same time fill up her water at a cascade on the north side of the bay. Our sportsmen, here obtained a sufficient number of sea-hens, which, when cooked in a sea-pie, made a mess for all hands.

A few days previous to our arrival at South Georgia, the ship, being at the time to the westward of, and between it and the New South Shetland Islands, was overtaken by a violent gale from the south, and hove to, and lay under storm stay-sails only; this was the winter season in this hemisphere, the weather also was extremely cold: while thus lying to, the ice from the spray of the short and breaking sea, made so fast, on her weather side, as to heel her over two or more streaks to windward, the violence of the gale to the contrary notwithstanding. This ice, so long as some expedient was not adopted to break it off, increased so rapidly, and withal so hard and flinty, as to bid fair to sink the ship; the utmost endeavors of the men with their handspikes to break it off, being found to be unavailing and fruitless. In this dilemma, the master was directed to wear, and lay her on the other tack, which being executed, of course gave her a very rank heel, but, as was expected would be the case, the sea water softened the ice, so that the men were able to break it off and shovel it over. To make use of every advantage, that the men might be enabled to hold out on this severe duty (the thermometer in an exposed situation being 36° below zero), they were divided into three watches, relieving each other every half hour: this was the more necessary, for while the severity of the gale continued, as fast as the ice was cleared from the lee side and deck, a mass of it would be collected on the weather side, so that every few hours we were obliged to wear ship; this was the only way in which, with the untiring and spirited exertions of the officers and men, she was kept from foundering.

Our excellent surgeon* was not unengaged while these duties were being attended to, for to his information, his advice, and strict attention, the officers

*Doctor G. Smith, of New York city; his kind disposition and attention to the sick and wounded, together with a regard for their welfare, endeared him to all on board, and the many happy and agreeable hours spent in his company during this voyage, makes it a pleasure for the author to bear this feeble testimony to his worth.

and men owe the preservation of their ears, fingers, toes, &c., for though many were severely frost bitten, yet all were cured by following his directions. One of his most prominent remedies, was to rub the part, immediately on being frost bitten, with snow, or soft spungy ice, then immersing it in ice water for a few minutes, or as long as the patient could well bear it.

An additional evidence had been now produced in favor of my former impression, that ice could not possibly make on the surface of sea water, be the frost ever so intense, unless there was some body, as land, rock, or vessel, for it to attach to in its commencement; as immediately after this gale, we had a calm and clear day; then not an ice island, bergh, or float, was to be seen as far as the eye could reach, except the pieces broken off from the ship.

We now steered to the eastward, and in a few days after, in the morning at day-break, Willis Island was in sight, bearing E. N. E.; shortly after, the black mountain peaks of South Georgia showed themselves above the masses of ice and snow. Under a whole sail breeze from the westward, we then passed to the north of Willis Island around Bird Island, and thence to the south-east along the coast. On arriving abreast of Sparrow Bay, where it was expected the crew of the *Regulator* would be found, we hove to, myself, with an officer and boat's crew, landing near the wreck of the R. Their habitation was found deserted, and from information afterwards obtained, it turned out than an English elephant oil ship had touched at this bay, at the close of the past season, to the captain of which the officers and crew of the *Regulator* had disposed of her cargo, consisting of rising 14,000 fur seal skins, with the sails, cables, anchors, rigging, &c., saved from the wreck, and after putting all the provisions on board, had taken passage in the same ship for home. Finding nothing left for the owners, we concluded to lay the ship by for the night, and early in the morning to proceed along the coast to Woodward harbor.

At daylight, bore up, passed the Bay of Islands, so called by Captain Cook, and at 10 A. M. were in the mouth of Woodward harbor, the wind being then light, and unsteady, mostly ahead. After furling the sails, we warped the ship into the harbor to an anchor. Here we met the English ship *Morse,* on board of which vessel was one of the *Regulator's* former crew, from whom the above particulars, in relation to that vessel, her officers and crew, were obtained.

The *Aspasia* now lay moored with three anchors ahead, and two out at the stern; this was necessary because of the gusts of wind which here whirl down the gullies, a westerly gale particularly, blowing directly out of the harbor, as it does, comes pouring down suddenly and heavily on a vessel, striking her first on one bow, then on the other, causing her to sheer and roll as much as though she was in a gale at sea; this was the case with our ship, notwithstanding her yards were all down, and the topmasts launched and housed. In the mountainous land which surrounds this harbor, there are a number of gullies (called gulches, by the seamen); it is by these the gusts of wind in the heavy gales come rushing one after the other, with a rapidity and force that forbid any attempt to look to windward, ofttimes throwing the water over us, as in a heavy storm at sea. Some idea of the same may be had from the fact, that the light cedar whale-boat moored at the stern of the ship, and held by the warp at her bows, has been taken up by these violent gusts, and turned over and over, before again striking the water, the same as a feather attached to a thread, and blowing in the wind.

Disappointed in not receiving assistance from the *Regulator's* crew, from whom much had been expected on our setting out, it became necessary to prepare for prosecuting our business with additional energy, as there was every reason to believe, that soon after the winter should break up, and the summer season set in, many vessels in the same pursuit would arrive at this place. As little can be done here in collecting fur seals without the aid of shallops, we set to work to supply ourselves with them, by first raising and decking our launch for one; then purchasing another, from an officer who had charge of her, and which had in the previous season been built here by the crew of an English elephant oil ship, then, taking a spare top-mast of the ship's for a keel, and a spare main yard for a mast, together with some fifty large oak knees roughly hewn, that had been put on board at New York, to dagger knee, or support our battery-deck, should the weight of metal placed thereon render this additional support necessary; with these materials, and a number of three inch oak plank, which had been used to floor the ballast over, and sundry articles procured from the wreck of the *Regulator,* we proceeded to lay the keel of the third vessel on an iceberg, in a valley at the starboard side of the harbor. The officers of the English oil ship made merry at the commencement of this undertaking, but still we had the gratification of answering their ridicule, by launching within the time of fifty days, a completely finished vessel of upwards of thirty tons. On trial, she proved to be a first rate sea boat, as well as the fastest sailer and best shallop among the fleet in the country.

As was anticipated, so it turned out; when the summer season set in, seventeen sail of sealing vessels, mostly ships, with their shallops, arrived at this island. We had rather the start, however, for our men having been previously placed at the different stations, and aided as they were by the fast sailing little vessel, were enabled, out of the 112,000 fur seal skins taken by the crews of all vessels during the season, to secure 57,000 for our share, a little over the half. By the activity and industry of the officers and men, these were procured, cured, and stowed on board ship, ready for sea, when the season closed. The vessel we had here built (by this time become quite celebrated) was sold to an English captain, for one hundred and twenty guineas, who very good naturedly remarked, that "Yankee enterprise and perseverance must ever stand pre-eminent."

While the greater portion of the ship's company were busy in procuring our cargo, a few (among whom were the surgeon, and at times myself), consisting of a midshipman and boat's crew, remained to keep the ship: these were frequently out on hunting and fowling excursions, to obtain fresh provisions, as well as to pick up what scattering fur seals might be met with on the same. On one of these excursions from the ship, the surgeon and myself had strolled, or rather climbed, to an elevated precipice of the mountain; from this position a wide and extensive view of the ocean was had, covered with fleets of immense ice-islands. The brightness of a clear sun shining on these islands, and on the sea as it broke against their base, formed a view, which for grandeur and beauty, is seldom if ever surpassed. The position we occupied was an acre or upwards of table level, and from this to descend to our starting point, was a performance by no means easy, in consequence of the steepness and slipperiness of the descent, for we had found, in this instance at least, that it was easier to go up than to come down hill; however, by exercising all due caution, we at last safely attained more comfortable quarters.

On the 19th of September, our lieutenant of marines, while prosecuting his favorite sport of fishing, was so fortunate as to discover a very welcome addition to our supply of fresh provisions. He had but just anchored his boat by the edge of the kelp, not dreaming of such great luck in store, when a fine cod, some eighteen inches or so in length, was taken by him; this was followed by his frequently catching them in pairs. A mess of these were sent to our neighbors, the captain and officers on board the English ship *Morse*, who were equally astonished with ourselves at finding cod-fish in these waters, never having dreamed thereof, and expressed themselves as much indebted for the present and information. These fish were found to be the largest in deep water, or the deeper the water the larger the fish; some were taken weighing between thirty and forty pounds. A car, made out of a large cask, was kept well filled with them, and enabled us to afford every mess a supply, when desired.

The ice islands, bergs, &c., formed during the severe winter frosts, and storms of snow and sleet in this high latitude, commence originally at the valleys between the mountains, which last also surround the bays, storm after storm, and snow drifts, congealed into solid ice, not only fills the valley to nearly the mountain top, but also makes out from the shore or beach at the head of the bay, until it extends as far as the capes, the outer edge making an inwardly curved line from one cape to the other, and spreading from mountain to mountain, over the entire surface of the bay, the thickness varying from a few, to many hundred feet. In the spring of the year, or rather summer, particularly in heavy rain storms, vast bodies of this ice crack off, across the entire width of the bay, being in size of greater or less magnitude, according to the extent and breadth of the bay. When these herculean bodies fall over into the sea, the noise is as if it were caused by the discharge of thousands of cannon at once, and apparently shake the earth to its very foundation; these detached portions afterwards again breaking into two or more pieces, are the ice islands, bergs, &c., met with in these seas.

When the water in the bay is sufficiently deep, which frequently is the case, they do not ground, but if the wind is off shore, drift out to sea immediately. In this manner, one mass breaks off after another, until the last breaks at the shore, so that the front rests on the land; the breaking of course then ceases, and leaves a perpendicular front to the main mass across the valley, of several hundred feet in height. On this, whenever the sun shines, various rich and brilliant colors are to be observed, giving to the main front a light blue shade. This body, however, gradually melts away during the summer, leaving from one to two-thirds remaining in the valley, when winter again sets in.

It is very advisable, that every commander who ventures into these regions, should be well acquainted with this formation, progression, and time of the breaking up of the ice, and also the formation of ice islands, for it would be, to any not possessed of such knowledge, extremely hazardous to venture into one of these bays during said seasons, as the risk of being instantly overwhelmed and destroyed, is exceedingly great. The packed ice islands, as they are termed, are several layers or flakes, piled one upon another; these are from the ice which makes first at the shore, thence spreading out to sea, a distance of several miles or leagues, in thicknesses of from one to two or more feet; the outer edge, the heaving of the sea in a heavy gale, by continually lifting and settling it, is broken off in different sizes, and as detached, is by the tossing waves, thrown upon the main and still firm or fast body, sliding far on it; as the gale, increases, flake succeeds flake; these, congealed by the frost, form an immense body, and after being piled up two, three, or more layers, their weight becomes so great, as to break off again from the firmer ice, and these are what are called the packed ice islands, which of course, when the wind changes, drift out to sea, and not unfrequently, are met with in portions of a mile in length.

A navigator, in endeavoring to make a passage among ice, should have much experience, and be exceedingly cautious in his progress, as such navigation is always attended with very foggy and thick weather. The first appearance of a change from the winter to the summer season at South Georgia, is discoverable in November; the ice then begins to break away, and the seals to come up; this is followed with an immediate destruction of their numbers by the sealers, with as much briskness as a due regard to the skins being kept well conditioned, will admit of.

On the 8th of February, 1801, the sealing season being over, our ship was made ready for sea; shortly after left her moorings, and was got under weigh to leave the harbor, but meeting a heavy swell from the sea at its mouth, and taken aback at the same time, we were obliged to clew all up, let go the sheet and best bower anchors in twenty-eight fathoms, to bring her up. In this narrow and dangerous position we passed the hours very uncomfortably, until the next morning, when the wind coming fresh from out the harbor, and from the land, relieved us; we instantly weighed anchor and were soon safely out to sea. After navigating between four or five days in the midst of islands, bergs, and floats of ice, accompanied with thick fogs and misty weather, by this means rendered additionally perilous, we at last attained a clear and open sea, crowding all sail to double Cape Horn.

February 25th. Sounded, and got ground in seventy fathoms on the bank to the northward of the Falkland Islands.

March 2d. Had sight of Cape St. John's, the most eastern promontory of Staten Land; passed the cape at meridian, and at 4 P. M. it bore N. N. W. distant ten leagues. In our passage, while doubling Cape Horn, we frequently had heavy gales, being mostly from the westward, and unpleasant weather, with snow squalls and sleet.

On the 16th, had sight of Cape Noire, the most southwestern cape of Terra del Fuego; after stretching a league to the westward of it, we tacked ship, and stood off to the S. W. On the outer rocky islets, we could distinctly see, with the aid of a glass, great numbers of seals.

April 7th. At 10 A. M. had sight of the Island of Mocha, bearing N. N. E. half E. about twelve leagues distant. Two days after, at 3 P. M. came to an anchor at the Island of St. Mary's, on the coast of Chili, there to wood and water, and dry our fur skins; at this place we found a small fleet of American sealers, being five ships and a schooner, from whom we learned there were upwards of thirty sail of American sealing vessels on this coast, whose cargoes were destined for the China market.

April 30th. Our sea stock of wood and water being completed, we weighed anchor, and laid the ship's course for Valparaiso, in order there to obtain refreshments, of which we were somewhat in need. At that place the ship was detained by the unjust proceedings of the governor, to whom the captains of the Spanish merchant ships then in port had presented a complaint, wherein it was stated they suspected the *Aspasia* to be an English cruiser. (England and Spain were at this date engaged in war.) Some of these worthies even went so far as to state their willingness to testify to it as a fact, being very sure we had two sets of papers, the American commission being merely a cover for our hostile character. However, after an investigation into the matter had been held, to the complete satisfaction of his Excellency, proving our ship entirely clear of charge, he apologized for the detention and trouble to which we had been put, and suffered us to depart in peace, which we accordingly did, after obtaining the desired refreshments.

On the 23d, we took our departure, and proceeded down the coast as far as Coquimbo, thence across the Pacific to China. We followed the old and beaten track of the Spanish galleons across the Pacific, and on the seventh of August 1801, at 9 A. M. had sight of part of the Phillipine Islands, being Tinian and Saypan, bearing five leagues distant. At this time, and for the twenty-four hours previous, large shoals of the sperm whales, black fish, and porpoises also, in great numbers, were spouting and gamboling about; in fact, the ocean appeared to be quite alive with them, and sundry other kinds of smaller fish.

On rounding the ship to under the lee of Saypan, it is difficult to conceive, unless by those who have been thus gratified, how greatly we were refreshed and delighted by the fragrance from off the beautiful green foliage of this island, after so lengthy a passage. At this island we lay off and on for several hours, waiting for the boat which had been sent ashore to obtain some

WHAMPOA, CHINA

From an oil painting by a Chinese artist, at the Peabody Museum, Salem

THE "CENTURION," CAPT. ANSON, TAKING A SPANISH GALLEON
From an engraving in Anson's *Voyage Round the World*, London, 1748

fresh meat; she shortly appeared, bringing along a large bullock which had been singled out from the herd, and shot for our especial accommodation, together with some cocoa-nuts and limes; these safely deposited on board, we bore up with a fresh trade for Canton, hove in sight of and passed the Bashee Islands September 2d, and on the 6th anchored in Macoa roads. The following day, after procuring a pilot and chop, weighed anchor for Wampoo, where we arrived on the eighth, and moored the ship.

The next morning I proceeded for Canton to complete the necessary agreements with a Hong merchant, secure the ship at the factory, and make all the arrangements for disposing of our present cargo and purchasing another, which the manner of conducting business with this people render requisite. Every thing was moving forward as usual; the same routine, the same regulations, not an alteration or improvement to be observed: the Chinese are a peculiar people in this respect, and tenaciously adhere to old customs and forms.

Some few days after, by a friendly invitation from the brother of the Hong merchant with whom our ship was secured, a mandarin of high order, the author visited the inside of the city of Canton, where but very few foreigners it is asserted are ever admitted. Our little party, consisting of the mandarin, myself, and the guard, made the best of our way through the suburbs, and arrived in good season at the city gate; here were some soldiers at post, who paid the mandarin military honors, as he passed in. Not many yards had we advanced, however, into the streets of the city, before a motley collection of all ages and sizes, from among the lower orders, attracted by the sight of so strange an animal as a Christian among them, began to follow on after us, keeping up a continued cry of *"Hi yah, Fonqui* (hallo Christian), how came you here?" which salutations receiving no response, were followed by bits of bones, dirt, rags, &c. thrown at the said unoffending Christian. Some of the adults were treated with a severe bambooing in return for this rudeness, a very speedy dispersion of their numbers following, so salutary a lesson on the leaders, as inflicted by the mandarin's orders to his body guard, the like treatment was not again experienced by us while we remained here.

The streets are mostly narrow, many not over twelve or twenty feet in width; some, viz. China street, being in general somewhat wider than those of the suburbs; they are, however, kept in a very cleanly condition. The houses were in most instances from one to two stories in heighth, built principally of brick, and covered with tile, much also like those of the suburbs. After passing several hours in visiting the public buildings, manufactories, &c., we returned through the same gate by which we had entered, without discovering any thing extraordinary or greatly different from what may be seen in the suburbs.

A circumstance tending to show the superstitious belief and attachment this people have in and for their God, Josh, took place some days after our cargo of sandal wood had been disposed of. On that day there lay at the ship's quarter a hawpoo-boat (the common term for a family boat), belonging to and on board of which was a very clever mandarin, from whom we had received many favors; he was at the time lounging in his parlor, but came out on the boat's deck on my calling him, and then asked what I wished. "There, sir," said I, handing a piece of red-heart hickory, taken from the lot our steward was splitting up, and which was afterwards hewn round in imitation of merchantble sandal wood, "is a *cum shaw* (a gift) of Josh wood for you." "Does it" he asked rather doubtingly, "have true Josh wood?" "Why, you have Josh man (you are a worshipper of Josh), you can *ser va* (know) that thing, I no can, not being a Josh man." He then turned the piece over and over, weighing it in his hands, not quite satisfied in his mind about the purity of the article, after smelling it again, he still doubtingly inquired, "Truly, does it have true (is it true) Josh wood?" "You have Josh man, and must *ser va* that," I replied. Again the close inspection, the weighing and smelling was renewed; yet there was no cheating him, for in a moment or two, shaking his head, he returned the piece, saying, "I much *chin chin* you, but truly he have no Josh wood." "Never mind, never mind, Josh won't know the difference; you keep it, and *chin ching* with it to him (sacrifice by fire), be assured it will answer." "*Hi yah*" (an exclamation of surprise and doubt), said he, as he turned to re-enter his cabin, "how can do that thing, and cheat Josh? suppose any man do such thing, Josh kill he at once." This sandal wood is kept constantly burning on the altars before their God, Josh, at the houses of worship: it is highly impregnated with essential oil, and when burning sends forth a strong fragrant perfume.

As soon as our return cargo was purchased and received on board we made preparation for sea, and on the 25th of October were under sail once more, bound to New York.

In the Borneo Sea, early in the morning, when, in company with a fourteen gun brig, three ships were at one time in sight from our deck; two of these very soon made sail and stood away, while the third, a large ship, which we imagined to be a French Indiaman in disguise, bore down towards us; this impression was greatly strengthened, by noticing as they came nearer, the red caps and shirts of the men, and the assumed broken English, in which the hailing officer demanded the name and destination of our vessel; his gruff hail, was retorted by a "what ship is that?" Of course, such a one sided affair as our mutual questioning, without any answer made it, was not calculated to throw much light on the subject; so that firing, as a more speedy mode of determining the question, was soon commenced. This was continued with but little damage to our side, until our opponent was observed to be opening a lower tier of guns, which it was quite impossible to have discovered before. This operation gave more information than an hour's firing could possibly have done, and without waiting until such fresh evidence of his strength was exerted upon us, a parley was had, and our united mistake explained. He also, as the boarding lieutenant informed me, had conceived our ship to be a Frenchman; so that, while thus attempting to take a British sixty-four gun ship (Lord Anson's old ship, the Centurion), mistaking her for a French Indiaman, we caught a

Scotch prize, or, as the common saying is, "burned our fingers a little." After this, we lost another of our midshipmen, Mr. C. Thompson, by the yellow-fever, on passing the Cape of Good Hope. He was a most exemplary young man, and much beloved by his brother officers.

March 4th, 1802. The *Aspasia* returned safely to New York: the close of this voyage, however, mortifying as it was, resulted far different from the *Betsey's;* this was principally occasioned by the deficiency in the company's receipts, as one of their vessels, the *Regulator,* had been wrecked, and with her valuable cargo being all uninsured, was of course a total loss; thus creating a gap which the *Aspasia'* voyage was not able to fill up, and afterwards leave a profit. Had this misfortune not occurred, the adventure would have been very profitable, and perfectly satisfactory to all concerned. This China cargo, arising out of a South Sea sealing voyage, paid also, a large amount of duties into the treasury of the United States.

CHAPTER XVI

VOYAGE OF THE BRIG "UNION"

THE brig *Union,* Captain Isaac Pendleton, was well fitted out, and amply supplied with the requirements for one of those expeditions, under the author's agency, similar to the voyages of the *Betsey* and *Aspasia.* Never, perhaps, was a voyage entered upon with brighter, and never did a vessel sail with more encouraging prospects than this brig. Her commander an upright man, and able navigator, was experienced, and possessed of every desirable capacity, successfully to prosecute this fishery and commercial trade. Mr. D. Wright, the first officer, had served out a term of apprenticeship to it, and was an excellent seaman and navigator, of much mechanical ingenuity, and quick penetration.

The celebrated voyages of Vancouver had just been obtained; in these mention was made that seals were numerous on the south-west coast of New Holland, but particularly of great numbers resorting to Seal Island, in King George the Third's Sound; in addition to these voyages, the manuscript of the discoverer of Crozett Island, was received. On the information contained in all these, Captain Pendleton was directed, in his instructions, to proceed by the way of, and to double the Cape of Good Hope for New Holland; it was farther recommended to Captain Pendleton to attain the situation the Crozett Islands were placed in by their discoverer, and in the event of his rediscovering, to give them a careful examination; being, nevertheless, left unrestricted, and at perfect liberty to act on all occasions as his judgment should direct, to make the most profitable voyage he could of it for his owners; taking care, however, to leave buried on Seal Island, a bottle, containing a letter, giving an account of his success, and observations relative to this coast, for the guidance and information of the commander of such vessel as the same company contemplated sending out, the following season.

In accordance with these instructions, after leaving New York and doubling the Cape of Good Hope, the brig's course was shaped for the Crozetts. After passing over their situation as laid down, with no prospect of finding them, she proceeded with all despatch for the coast of New Holland. On making the south-west cape of this coast, she ran it down a mile or more from its shore, until arrived at King George the Third's Sound, where was found a good harbor, wooding, and watering. At Vancouver's Seal Island, although it was the period in the season to expect the seals would be up in great numbers, there was nevertheless not above thirty on the island. Here, after remaining some time, without any others arriving to increase their numbers, and having discovered none of these animals, while tracing to the westward, it was concluded most advisable to proceed and examine the coast to the eastward. After surveying this a considerable distance, the *Union* was overtaken by a heavy gale, and driven to sea before it abated, some degrees to the S. S. E. from the coast; when the wind began to moderate, they laid their course to the north-east, and in latitude 35° 47' south, longitude 136° 41' east of Paris, discovered Borders Island. On this were found both the hair and fur seals, extensive forests, good water, and much game; fowls and birds of various kinds in abundance; and also excellent fish and oysters in great plenty: there are at this island two harbors, one, on the north-east part, in latitude 35° 40' south.

In this the brig lay snugly moored, while her company commenced taking the seal, which was continued for several weeks with considerable success; but as the chief part of the season had gone by for sealing, before the discovery of the island had been made, they were only able, before it closed, to procure the moiety of their cargo. In order to take every advantage of things as they were situated, it was decided to build a small vessel, in which Mr. Wright might proceed to King George the Third's Sound, there to deposit an account of proceedings, according to understanding, and make farther examination for seals. For such a purpose, the forests afforded plenty of very excellent timber; this, by the ingenuity of the first officer, assisted by the carpenter and armorer, was hewn and sawed into planks, as needed; so that with these, and such materials as were had from the brig, a substantial schooner of forty tons burthen, was very shortly launched into her element, sails and rigging being on board the *Union* in sufficient quantity.

On the return of this schooner to Union harbor in Borders Island, from her trip to King George the Third's Sound, where no farther encouragement respecting its being good sealing ground had been obtained, than on the first visit, Captain Pendleton concluded to proceed to Port Jackson, New South Wales, from which place both vessels, after obtaining an addition to their provisions and water, were to start in search of islands placed on some ancient charts, and said to have been discovered by Tasman, with other early navigators; thus improving the time until the sealing season in this hemisphere should return.

Accordingly the *Union* sailed from Borders Island, accompanied by her tender, bound for the western mouth of Basis Straits, through which, unaided by charts or directions, he found his way, without the slightest accident, being, as is believed, the first American vessel which had passed through, and arrived at Sidney, near Port Jackson, whence completely refreshed and recruited, he sailed with the intention of examining the southern coast of Van Diemen's Land, as well as to make what discoveries he could of other islands.

Towards the conclusion of a lengthened cruise, and about the setting in of the sealing season, he had the good fortune to rediscover the Island of South Antipodes: here they found large rookeries of fur seals, but finding no harbor in which to moor the vessels, after leaving an officer and eleven men thereon, to take the seals and cure their skins, well provisioned, he was content to make the best of his way back to New South Wales. Notwithstanding the very flattering prospects now before him of making a most splendid voyage, increased almost to a certainty, from the determination of the men left at the island to have a full cargo of first quality skins, ready by the time he should return, he was induced by some temptation laid before him, by a Mr. Lord, merchant at Sydney, to discharge those skins already on board, amounting to 14,000, collected at Borders Island, and place them under the care of, and in said Lord's store. An agreement was then entered into between them, by which the brig was to proceed on joint account for the Fee Jee Islands, to procure a cargo of sandal wood for the Canton market, the merchant furnishing the requisite means for trading, by which the wood was to be purchased.

In pursuance of this new engagement, the *Union* sailed from Sydney, touching first at the Island of New Amsterdam, or Tongataboo, to engage a native, speaking the Fee Jee language, to proceed with them, to act as interpreter. Arrived there, the captain, accompanied by Mr. J. Boston, agent for Mr. Lord, and a boat's crew, were all most inhumanly massacred, soon after landing.

On the succeeding day (October 2d, 1804), Captain Pendleton not yet returning, Mr. Wright, the then commanding officer, became uneasy and alarmed; his suspicions and fears were farther increased, by observing that the chiefs and natives, while unhesitatingly declining to come on board, endeavored by signs to convince him that the captain wished another boat and crew to be sent to the shore, in order to bring off the hogs, &c., which were there said to be collected for them. In answer to all these, Mr. Wright gave them to understand, that the captain must first return, and they keep farther off, and desist from coming any nearer to the vessel. The canoes had by this time become very numerous around the brig, and were all well filled with warriors, gradually getting, nearer and nearer, until finally the natives became so bold as to close up alongside, many of their numbers armed with clubs, and other implements of war, clambering up the sides and into the channels; but noticing the preparations going forward on deck, for immediate and stout defense, and the charging of cannon, &c.; while Mr. Wright once more beckoned them off, as pointing to the guns, the threat to fire upon them, in case of nonobedience, was repeated; they left the brig and drew off.

At this moment another canoe was observed advancing very quickly from the shore, and to their surprise containing with the others a white woman, standing at the bows; after passing through, and gaining the space within the circle formed by other canoes, and getting nearer the brig, the chief the while drawing all attention to himself, from the earnestness of his manner, as addressing the woman, she suddenly sprang overboard towards the vessel, and upon rising again to the surface, in few words informed Mr. Wright that the captain had been murdered, and the farther intention there was to capture the brig. To deter the natives from pursuing and overtaking her, a volley of musketry was discharged between them; this had the desired effect for a moment, but no sooner had the noise and their surprise subsided, than a most determined attack was commenced; far now from being frightened at the rapid and heavy discharges of cannon and musketry, encouraged too by a great increase to their numbers, they bade defiance to every thing; the sinking of a canoe, or scattering amidst the survivors, of the mutilated limbs of others, the groanings of the dying, and floating bodies of some more spirited than prudent of their party, were regarded, if at all, but for a moment; a fiendlike determination seemed to goad them on, and a settled resolution, cost what it might, to capture the *Union*. As a most speedy termination to the battle, Mr. Wright ordered the cables to be cut, and sail to be made; this was the more advisable, as a farther sacrifice of their lives would scarcely have been justifiable. Undeterred they still continued the fight, shouting and hooting, until the vessel had left the harbor, and was out to sea; they then ceased their exertions, and returned to the shore.

The woman then proceeded in giving some account of what had come under her observation, on shore, by stating, although the plans were well matured, and the minds of the chiefs made up to cut the brig off, it was politically concluded as best to receive the strangers under the guise of friendship; this appearance was well carried out, the captain, Mr. Boston, and the crew, after landing being conducted over a hill, and entirely out of sight of those on board: here, in the midst of a parcel of trees, and surrounded by immense numbers of natives, they were all massacred. After this, the plan for completing their enterprise, was to request, another boat might be sent for the hogs, roots, and fruit, said to be awaiting them at the shore, whose crew were to be similarly treated. Wishing to escape and save the remainder of the crew, this woman had succeeded in persuading the chiefs to believe, that by

TONGATABOO, TONGA ISLANDS

From an engraving in Wilkes' *U. S. Exploring Expedition*, Volume III, Philadelphia, 1845

CAPE OF GOOD HOPE AND TABLE MOUNTAIN
From a drawing by H. C. DeMillion

taking her along in the canoe with those who were to make the above request, she could induce the commanding officer to consent and send another boat; which resulted as has been already detailed.

Eliza Mosey (as she stated her name to be), had arrived at this island in a ship called the *Duke of Portland,* commanded by Captain L. Melon; instigated by the cruel advice of two renegadoes, a white man named Doyle, and a Malay previously left on the island, who were far greater savages, and more blood thirsty than the natives themselves, these last had been induced to cut the ship off, and massacre all her company, except Eliza, a colored woman, a decrepid old man, and four boys, whose lives Doyle had spared, that they might assist him in discharging vessels, afterwards in burning and tearing them to pieces, with the double intention of securing the iron, and preventing any possibility of detection, or causing alarm to such vessels as might chance to stop for provisions and refreshments. To secure the capture of these as they should successively arrive, as well as to gain favor with the leading chiefs, and gratify their savage dispositions, Doyle continually presented to their view the quantity of fire arms, iron, and goods suitable for trading with the other islands, each of these vessels most generally had on board; thus prompted and instigated, they were ready and anxious, by duplicity or force, as would most likely succeed, to enter into his diabolical plans. Providence, however, did not long suffer this wretch to survive the capture of the *Duke of Portland;* the old man formerly mentioned, with the four boys, and a few natives were every day engaged, under his own immediate superintendence, in discharging the ship's cargo into the canoes and boats; this had been continued for a few days, and still but little progress made, when the old man and boys, who had been watching a suitable opportunity, embraced the moment when his attention was otherwise engaged, and destroyed him; afterwards driving the natives overboard, and immediately cutting the ship's cables, made sail (for these had not yet been unbent), and stood out to sea; this (after repeated inquiries on the part of the author, to ascertain what subsequently befell her), was the last account ever obtained, concerning this unfortunate ship and her crew.

Thus in command, Captain Wright determined to return to Sydney, in order to replace the number of the crew recently destroyed, and this being accomplished, the *Union* again sailed for the Fee Jee Islands, to fulfil the contract entered into with Mr. Lord. On arriving near these islands, they experienced a heavy squall, during which the brig, being a full-built, dull vessel, got embayed among the coral reefs, and in the calm immediately succeeding the squall, by the force of the sea swell and current, was cast on the reef and wrecked. Every person on board either perished by drowning, or was massacred by the natives, who from the commencement of this series of disasters had been watching every movement, and as each unfortunate man gained a foothold on the rocks, thus terminated his existence; their bodies, as has been subsequently ascertained, serving the purpose of food, for it will be remembered these islanders are cannibals.

Upon the arrival of this sad information at Sydney, Mr. Lord chartered a ship and proceeded with her to the Island of Antipodes. At this place, the officers and crew whom Captain Pendleton had left, had taken and cured rising of sixty thousand prime fur seal skins, a parcel of very superior quality; these, from information since obtained, were received on board Mr. Lord's ship, who thence proceeded to Canton and disposed of his valuable cargo at good prices, the proceeds being invested in China goods, he accompanied to an eastern port in the United States; these were also sold, and Mr. Lord made off to Europe with the amount of proceeds, before the agent for the owners of the *Union* was made acquainted with the transaction; thus unfortunately terminating the *Union's* voyage, her owners never receiving either for the skins taken from South Antipodes, or for the fourteen thousand left by Captain Pendleton in Mr. Lord's charge at Sydney, one farthing. Nor was the remainder of the brig's company more fortunate than their messmates, for nothing was ever heard of the few who after delivering the skins to Mr. Lord embarked on board the little schooner and sailed for Sydney, in New South Wales; it is supposed they were either lost in a heavy gale at sea, or were wrecked on some unknown reef or island. Thus terminated a voyage than which, none was ever commenced with more encouraging prospects, and thus went her crew, than whom, more hardy and resolute spirits never strode a vessel's deck. Notwithstanding this treatment to the *Union's* crew, the ship *Hope,* the next vessel sent out under the author's agency, succeeded in procuring a cargo of sandal wood at the Fee Jee Islands, the same having been taken to

Canton and exchanged for China goods, with several other cargoes of the same article, as also others of beach la mer, mother-of-pearl, turtle shell, &c., have paid duties thereon to a very considerable amount into the United States' treasury.

Under the impression that Captain Pendleton and his men were detained as prisoners by the king and chiefs, and with a view to obtain their release, orders, from which the following is an extract, were given to Captain R. Brumley, of the ship *Hope*, (for at the date of the *Hope's* departure from New York, the sad particulars in reference to the *Union's* crew were not yet known to her owners), viz.

Sir,—You are desired to take charge of the ship *Hope*, now at anchor in this port, and put to sea the first fair wind, proceeding by way of the Cape of Good Hope, round New Holland as speedily as possible, for the island of Tongataboo, one of the group called the Friendly Islands, in the Pacific Ocean. Arrived at Tongataboo, you will proceed round Van Diemen's Point, at the west end or side of the said island, and enter Van Diemen's Bay, at the head of which is Adventure Harbor, wherein you will anchor your ship, and use your utmost endeavors to get off on board your ship, Captain I. Pendleton, and his men, who some time past were taken at the moment of landing with a boat's crew, and detained by the natives at this harbor, he then being in command of the brig *Union* of this port. We recommend to you, as the best method to deliver Captain Pendleton and men, and have them into your hands, the following, viz. After you have brought your ship to anchor in this harbor, keeping prepared for defense and safety, endeavor, by persuasion, presents, &c., to get their king, or two or three chiefs into your power on board, then give them to understand that to get their liberty again, they must send forthwith an order on shore, and bring off to you Captain Pendleton and his boat's crew, or, that all the white men on the island must be brought to you for the king and chief's ransom. You will consider yourself freely at liberty to remain in this harbor a sufficient time to put into execution all the means and plans in your power necessary to obtain Captain Pendleton and his men; and in our anxiety to accomplish this, we feel it a duty again to request that you will therefore not leave any means in your power untried, to release Captain Pendleton and his companions, that you may have the inexpressible pleasure of bringing and returning them to their families and friends. We also recommend to, and authorize you, a few days previous to your arrival at Tongataboo, to inform your officers and men of the object of the voyage, at the same time give to each a written certificate wherein you will engage to pay unto each officer and mariner, an extra pay, provided they will promise and engage to exert themselves to their utmost when at the island, in assisting you to release Captain Pendleton and men.

Respectfully,
E. FANNING, Agent.

New York.

CHAPTER XVII

VOYAGE OF THE SHIP "CATHARINE"

THE season succeeding the sailing of the brig *Union* from New York, the ship *Catharine*, a fast sailing vessel and excellent sea boat, was purchased, armed, and provisioned, with everything requisite for a two and a half years' voyage; her company was well selected, and were all able officers and seamen, the whole under the command of Captain Henry Fanning, brother to the author.

Great hopes were entertained that this vessel would re-discover the Crozett Islands; a copy of the discoverer's manuscript, of their situation, &c. was furnished Captain Fanning, who, by his instructions, was directed in case of not meeting with success in taking seals on the coast of New Holland, to exert his utmost endeavors to re-discover them, even at the expense of two seasons' cruising, for it was scarcely possible to persuade one out of the belief in their existence, their discoverer had been so very particular in his description of them.

All duly prepared, the *Catharine* sailed from the port of New York, bound for New Holland, by the way of the Cape of Good Hope, solely to endeavor on the passage out, to get sight of the Crozetts, and not to delay in searching for them, for hopes were yet sanguine that a cargo of fur skins for the China market, could be readily procured on the coast of New Holland. Captain Fanning was also furnished with the necessary directions to find the letter Captain Pendleton, of the *Union*, had deposited on Seal Island, at King George the Third's Sound. On their arrival there, after passing over the given situation of the Crozetts, without having obtained the least trace of them, the *Catharine* was brought to anchor. Meanwhile, the captain proceeded in the boat for Seal Island, there to search for the deposited letter; this he easily found, but was much disappointed on perusing its contents, to learn there was not the slightest prospect of procuring a cargo of fur seal skins on this coast. However, to make the best of the matter, he concluded that his best policy would be to return to his officers and crew, and undiscouraged in appearance to set before them, that now there was no other course left to be pursued than to search out and discover new hunting grounds. He therefore, on his return, had all hands mustered on deck, and stated the contents of the letter, and the improbability of their succeeding here, desiring them, at the same time not to be discouraged, as from certain information he possessed, he was confident a different result awaited them at the islands in the Indian Ocean, to the westward of where they then were, for which, after recruiting and refreshing themselves, they would make the best of their way; not one dissented, but all cheerfully consented to stand by their commander to the last.

During the *Catharine*'s stay in this Sound, the natives unreservedly came to the officers and men at their different stations on shore while employed in the wooding and watering business, and readily assisted the men in bringing the wood to the beach, but could not refrain from indulging in their thievish disposition on every occasion that presented itself, immediately thereupon starting for the woods and high grass; at first, a musket discharged over their heads frightened and induced them to return and restore the stolen article to the officer; but soon finding the discharges did them no harm, they refused to give them any notice, while their companions, who were standing among the ship's crew in the best humor imaginable, seemed scarcely to know that anything wrong had occurred. Unwilling to harm these, the captain gave special directions for every one to keep on the alert, and if possible, prevent any temptation from falling in their way. They were looked upon, by those on board the *Catharine,* as in fact the most miserable of human beings: it was scarcely possible to conceive the wretchedness of their condition, having no settled residences, they were constantly wandering about from one place to another, and were, as the crew called them, a species of one half human, the other belonging to the baboon.

Two of the chiefs received an invitation from the captain to take breakfast on board the ship; he, having previously observed that his guests were remarkably fond of fried fish, had plenty of this prepared, together with a suitable quantity of coffee, bread, &c. for their own, more particular gratification. The invited guests were placed at the after end of the cabin table, himself and officers seating themselves around, before each of the former, the steward, according to directions, had placed a goodly quantity of the fish. Knives and spoons conveying too dainty a morsel for their liking, were left unemployed, hands the while performing the duty, and stuffing as much into their mouths as could there be crammed; as if fearful there was no more for them, another small lot was thrust in by way of filling up, the whole then being twisted and turned about so that the bones might work or be picked out at the corners of the mouth. In the effort to swallow such a tremendous portion, it was necessary to stretch the neck a little and bring the head forward, a performance somewhat like that acted out by our domestic fowls, who good naturedly have almost choked themselves with Indian meal, the execution requiring sundry laborious attempts to swallow the mass; when this was happily achieved, another mouthful was made to follow as speedily as might be. One of the chiefs having his mouth thus comfortably filled, pointed to a dish of brown sugar, and the captain supposing he had set his affections upon having some, took a small matter of it in a teaspoon, and as well as he conveniently could, without being rude, put a little in this chief's mouth, along side of what was already there lodged, some crowding being necessary to do this however. One would hardly have thought he could taste it; but he did, and not at all liking it, gave one puff, and very unceremoniously blew fish and sugar, pell mell, over the dishes and table; the remains, of the sugar, which had sought shelter behind his teeth, or elsewhere, being ejected in the same way. This the officers thought was rather too impolite, and rose to leave the table; the captain, however, reminded them, it was best not to regard the offense, lest the invited guests in turn would consider themselves offended. After calling to the steward for some other bread, fish, &c., they reseated themselves, and concluded the repast in good humor. As to the chief, he had immediately replenished his mouth, carefully avoiding any more sugar, not waiting to be twice asked either, so to do.

The game found at the sound while the *Catharine* was there, consisted of the kangaroo, black swans, ducks of several kinds, plover, &c. together with a great plenty of excellent fish and oysters. After laying in a good sea stock of wood and water, and recruiting their stock of provisions, the ship was got under way, and worked back on their outward track, against the continually prevailing westerly winds. On crossing the supposed situation of the Crozetts, many signs of land, such as seals, birds, drift, &c., were seen; these added strength to the belief that the islands were to be found, and after proceeding to Prince Edward's Islands, were an officer and crew with provisions, were left to take seals, they returned, and continued cruising for the Crozetts, until the season was out, equally unsuccessful as formerly, except some additional signs of land.

After this fresh failure, the *Catharine* sailed for the Cape of Good Hope, there to pass a few weeks of the winter, while its severity should continue; there the vessel underwent a thorough overhauling, thus being placed in the best possible condition; whence, with provisions and stores recruited, she sailed to the south again, first touching at Prince Edward's Islands, to inquire after their companions, and leave them some of the Cape refreshments. These were found to have been very successful, which was a kind of stimulant to those on board, and tended to renew their spirits for greater exertions. After leaving their shipmates, a more thorough and extensive search for the Crozetts was had, but still several weeks were passed before, as working the ship much farther to the westward, after an extensive stretch to the southward, and going about, at midnight, on the other tack, they at daybreak discovered the southernmost island, bearing E. by N. three or four leagues distant. With joy and gladness of heart, they immediately bore up for it, and run down along its south shore; on doubling round the south-east point, they opened a bay, on the shore of which thousands of fur seals were discovered, while at its head was a basin of a harbor, with a bar of kelp across its mouth, over which a vessel drawing not more than ten feet water, can easily pass: this proved to be the only harbor these islands afforded. At this they landed, and as Crozett, the discoverer, did not so do, there is no reason to doubt but that these American citizens were the first human beings who had ever stepped on its shores. This, the southernmost, they called New York Island, and soon after rediscovering the other two, called the westernmost, a long low one, Fanning's Island, and the easternmost, a high and mountainous island, they called Grand Crozett. Although these were found to be situated more than one hundred miles to the south of the latitude as laid down on the ancient charts, there can be no doubt, the description and bearings from each other agreeing with that therein given of them, but that they are the islands Crozett discovered.

Having thus, finally succeeded in finding the object of their long cruise, on which there was a vast many fur seals, their next object was to return to Prince Edward's Island, for that portion of the crew left there. In accordance with these intents, so soon as the boat returned from her examination, they made sail from the bay, which had received the name of Catharine Bay, taking along, as fresh provisions, a few young seals and proceeded to the westward, passing close to Fanning's Island. Arrived at Prince Edward's Islands, the fur seal skins there collected by their shipmates were immediately taken on board, together with whatever belonged to the men; waiting, after this was accomplished, a day or two, that the suspicions of their having discovered any island might not be excited in the minds of officers and men, belonging to two vessels, one a ship, Captain Percival of Boston, the other from Hudson, who were there stationed on like duties, as their own had been. It was also a part of Captain Fanning's instructions, in the event of his succeeding in finding the Crozetts, and there being seals on them, to deposit on the north-east point of this island a letter, giving their true situation and bearings from Prince Edward's Islands, for such vessel or vessels as might next be sent out. To accomplish this, unperceived by the individuals belonging to the other vessels, it was necessary to exercise some caution, but having effected it, their tents were struck and all embarked. Near to the end of the point just spoken of, a heap of stones was snugly piled up, some two or more feet in height, while thirty feet southwest, by compass, from its centre, two feet under ground, a bottle containing the letter was placed, and that no evidence of its being there might be observed, the earth and turf were carefully replaced. Sometime after the *Catharine* left, this pile was discovered by those then left on the island, and suspecting some information there lay buried, which would be for their advantage to obtain, they set to work, removed the heap, and dug a hole several feet deep, and as many in diameter; in return for this "labor lost," going away as wise as they came, supposing it to be only a hoax upon them.

A few months after, these men still remaining, the ship for whom the letter was intended, arrived at the island; when her commander despatched the supercargo, with an officer and three men, to look for it. Observing that this party as they landed, had a compass, spade, and ball of line attached to a seal club with them, and start for the point, these men came up, and very readily proceeded to tell all they knew about the matter, by saying, "we know that you are after finding a heap of stones, but save yourselves the trouble, and leave your spade behind, for we have already removed the pile, and dug a hole large enough for a well, under where the heap was; so you'll find nothing there." Being asked to show the way, they consented, and walked down to the point; the supercargo, nevertheless, would fain try; and so setting down the compass, and placing the seal club erect in the center of the hole, he stretched out the line to the south-west, and then removing the turf, uncovered and took out the bottle. Seeing the letter distinctly therein, these strangers begged the supercargo to read it for them;

but this could not be attempted, the direction being for the commandant, who, on receiving it, got instantly under way, and left the inquisitive people not much enlightened as to his destination.

On arriving at the Crozetts, the *Catharine* landed her first officer with a gang of sealers, who had volunteered at Basin harbor, New York Island, to procure a cargo of seal skins, and have the same ready for the vessel belonging to the company, under the author's agency, which should next arrive there, as the *Catharine's* quantity of skins was soon prepared and safely stowed on board, the ship was shortly after making the best of her way for the Straits of Sunda; but his cargo being in a green state, in the salt, Captain Fanning thought proper, on his arrival there, to moor his ship in the harbor at New Island, the same being uninhabited, and there dry and cure his skins, in the best state for the China market.

A sufficiently large space on the rising ground adjoining the harbor, and conveniently at hand, was cleared for the purpose. They had only been a few days thus engaged, when a small proa made her appearance off the mouth of the harbor, reconnoitering them; and her not coming alongside, when the American colors were set upon the ship, caused the captain to suspect her intentions were not honest or friendly. He, at first sight, supposed she had come from Anjier point; but her manœuvring soon induced him to change this opinion. An officer with a glass was accordingly posted in a tree on the elevated ground, to watch her motions; where, after reconnoitering a little while, she was observed to pass round the east chop of the harbor, steering directly across the strait for Sumatra, and this course was keeping when the officer could no longer distinctly see her: upon this report, the captain felt assured she could be none other than a scout from the Malay pirates, and caused him to suspect an attack would be made upon his vessel, before their business should be completely finished. It was, therefore, prudent to prepare for the worst; to do this the more effectually, a masked redoubt was quickly erected on a small hillock, overlooking and having the entire command of the harbor and its entrance; therein were placed two of his six pound cannon, in charge of an officer, and eight of his best men; these were every night on guard duty. In addition to this defense, the remainder of the ship's company was required to be on board at night fall, and these again were divided into three watches, alternately relieving each other during the night. Thus, with arms in the most serviceable order, and prepared for action at a moment's notice, a sentry too, posted in the tree formerly spoken of to keep watch by day, they made the most of their time; and had great reason to be very thankful, before the whole cargo was dried, that these precautions had been taken; for early in the day, a fleet of piratical proas from Sumatra, were to be seen crossing the strait directly towards their position, keeping close under the shore (evidently supposing themselves unobserved); here the advance arriving near the east point of the harbor, made a halt until the remainder of the fleet, who were somewhat in the rear should come up. So soon as all were joined, the whole party pulled away merrily at their sweeps, and while entering the harbor, set up a loud shout, making right for the ship; she having a spring on, veered her cable, so as to bring her whole broadside to bear, at the same time giving the signal (for it was distinctly understood, that the party in the redoubt were to remain in close ambush until this was given from the vessel, the better to secure a more decided result from their united fire) to the battery; the two at the same moment commencing the engagement, somewhat staggered the pirates in their advance; the reception was quite unexpected; they instantly backed water, and making a sudden stop, seemed cogitating what was best to be done; meanwhile, a merry peal was keeping up by the fort and ship, disabling some of the proas, and damaging others; their commodore, however, shortly made the signal for a retreat, adopting the resolution, that "prudence is the better part of valor;" whereupon their whole fleet put about, and pulled briskly for the opposite shore from the fort, taking in tow all the disabled proas, with the exception of one; the ship's launch was immediately manned and sent in pursuit of this, and succeeded in capturing her, before she had got out of the reach of the battery guns. After taking away all their armament, and endeavoring to enlighten them by signs, &c., as well as he could, never again to attack a ship wearing their colors, as one half of their crews were always awake and prepared for action, the captain sent them about their business, with a promise, which they little regarded, to take every man's head belonging to their fraternity, who should at any future time make an attack on his vessel, and thus become his prisoners. The piratical squadron, after doubling the point towards which they were sailing, directed their course close in under its shore, until fairly out of the reach of the battery guns, and then proceeded across the strait for Sumatra. Thus ended this affair, and although not again troubled for the few weeks, it was necessary for them here to remain; still the watch and guard kept as strict a lookout as formerly.

A few days after this brush, an officer from the Dutch station at Anjier point, came to the ship in a boat, and stated that the natives of the Java shore, not far from the Mew Islands, who had been at the Anjier market, and had given information of an attack, made by the Malay piratical proas, upon some ship, which had beaten them off, as after some heavy firing the pirates were seen recrossing the straits for Sumatra, his visit being for the express purpose of purchasing as slaves the prisoners, which he very reasonably supposed the ship had taken. He was a little chagrined at thus getting only his trouble for his pains, and said, as the captain told him they had been only disarmed and then sent away, it was a pity he had treated them so easily for being a blood-thirsty set of villains, they would soon be cutting off some other ship, probably his own countrymen, and murder her crew; advising the captain, if he should be again attacked, and take any more prisoners, to bring them along to Anjier, which lay on his route for Canton, where they were much wanted, and would readily bring three hundred Spanish dollars a head; and as he thus concluded, taking his leave.

The ship's company after this pursued their business undisturbed, and finally completing the relading of the ship, with wooding and watering, made sail for Canton, where her cargo of skins was exchanged for such goods as suited the New York market. To this last place she returned after an ordinary passage; netting, on the closing up of her accounts, a very handsome profit for her owners.

CHAPTER XVIII

SEA ELEPHANTS AND FUR SEALS

THE SEA ELEPHANT. These amphibious animals, at the proper season, come up out of the sea in various numbers at a time, and on reaching the beach, lay in rows along the same, such being what are technically termed rookeries, though some contain many more than others. The full aged males alone have the proboscis, and some of these are truly enormous animals, varying from twelve to twenty-four feet in length, with a proportionate height and breadth. The females, at this season, come on shore to shed their coat(as do also the males), and bring forth their young; they have generally one, sometimes two, never more at a birth, and rarely, if ever, even at full age do they attain over half the size and dimensions of the male.

On land, the elephant is a very loggy (a sea term, meaning heavy in their movements) animal and except among themselves, or in their own defense, never make battle. They are taken for their oil, and tongues, which are considered a delicious dish, and more luscious than neat's tongue. When first coming to their favorite shore (a sandy or pebbly beach), the animal is exceedingly plump, and very fat, the full grown generally yielding about three barrels of oil; but in a few weeks it falls away, becomes lank and poor, and by the time to go off comes about, would not give above the half of that quantity. In taking the younger, a club is commonly used, and for the old ones, a lance; yet in order to overcome the largest bulls, it is necessary to have a musket loaded with a brace of balls; with this, advancing in front of the animal, to within a few paces, they will rise on the fore legs or flippers, and at the same time open the mouth widely to send forth one of their loud roars; this is the moment to discharge the balls through the roof of the upper jaw into the brains, whereupon the creature falls forward, either killed, or so much stunned as to give the sealer sufficient time to complete its destruction with the lance. They are frequently discovered sleeping, in which case the muzzle of the piece is held close to the head, and discharged into the brain. The loudest noise will not awaken these animals when sleeping, as it is not unusual, though it may appear singular, for the hunter to go on and shoot one without awaking those alongside of it, and in this way proceed through the whole rookery, shooting and lancing as many as are wanted. The quantity of blood in these animals is really astonishing, exceeding, in the opinion of the author, double the quantity found in a bullock of the same weight; when killed, the whole thickness of the blubber or fat, with its skin, is cut into strips of from five to ten inches in width, according to the animal's size; and these cut from head to tail, torn from the carcass and separated from the lean flesh, are then, washed clear of the blood and taken to a mincing table, where the skin is taken off, while the blubber, after being cut into

A MALE AND FEMALE SEA ELEPHANT

From an engraving in Anson's *Voyage Round the World*, London, 1748

SEAL ROOKERY, BEAUCHENE ISLAND, FALKLAND ISLANDS
From a lithograph in Fanning's *Voyages*, New York, 1833

pieces about two inches or less in size, is thrown into a kettle and tryed out, the oil thus produced being put into casks, the scraps always furnishing plenty of fuel for the try kettle; a new cask after being filled with the boiling oil, is then started and coopered, necessarily, from not being fully shrunk, requiring to be filled again with the boiling oil, and even a third time, if it has not done shrinking after the second filling, which can readily be discovered; this course being particularly attended to, it may finally be coopered and stowed away in the ship's hold, to be filled up by the hose, and will remain tight for the voyage, in all climes, nor require wetting for any length of time, or lose a gill of oil by shrinkage. This it is presumed would be the case as to the casks with sperm or other oils; the author is also of opinion that without this careful method, a much greater loss will take place from leakage than is experienced by this process of shrinking the casks with the boiling oil, even though the casks are frequently wet, to do which in a ship's hold at sea in any weather, is always an unpleasant job, and requires a vast deal of labor. In the maws of the sea elephant, a quantity of gravel or sand is generally found.

The Sea Leopard. This amphibious creature, differing in several respects from the sea elephant, although in size it is nearly as large, is very smooth and neatly built; the fore paws or flippers are shorter, and on shore they are a more clumsy animal than the elephant; the hair is short, sleek, and spotted, as the land leopard. If frightened or disturbed in any way, they commonly tumble and roll down a steep beach (when there) for the water, with surprising quickness, and when in the sea can move about very nimbly, and in point of swift swimming, are equal to the herring hog, cutting their way through smoothly and rapidly; there they are more at home than on land, and are more courageous and daring, often following the boats with the seeming intention of attacking them, and this it is said has occurred.

Like the elephant, they are covered with a thick slab of fat; the skin, however, is thinner, and not so tough. They too observe the seasons for coming up and going off in the same manner, but are not with the elephant an inhabitant of all climes, as it is the author's opinion that the sea leopard never has been seen under the fiftieth degree of latitude, or even down near to it. They are but seldom found at Staten Land, the Falkland Islands, New Zealand, Desolation Island, or South Georgia, and there, very scatteringly, or only singly and in pairs. At New South Shetlands they are more numerous, not so much so even there, as to herd in rookeries; but on Palmer's Land, and the south part of Sandwich Land, they are found herded together in rookeries of many hundreds, and furnish oil, as the elephant, in proportion to their size.

The Sea Lion. Also an amphibious animal, is the full-aged male hair seal; the female of the same species, is called a clapmatch; the half-grown yearlings; and those less than a year old, the pups. The legs or rather flippers, of the lion, are longer, and the animal itself much more active on land than the elephant. Of this species of hair seals there are two grades or kinds; one much larger than the other, and having longer hair; it is of a brownish color, is also the most numerous and ferocious, and is an inhabitant of both the high and low latitudes. The other kind is a handsomer and neater formed animal, with short and sleek hair of a russet or light sorrel color, with a round and long neck; this last description the author has not known to be taken other than at the Gallipagoes, and the islands to the north of them on the coast of Mexico. The first mentioned grade vary from nine to fifteen feet in length, very large around the neck, but tapering off gradually till quite slender at their hind flippers, with a head resembling the land lion, good sized white teeth, and a very long bristling mane. On approaching near to them when the clapmatches have their young, which is one at a time, and for which purpose they come up on shore, they immediately advance to their defense, shaking their manes, roaring and growling and showing their fine teeth, and presenting to a person unacquainted with them, a terrible and powerful appearance. They are taken in the same way as the large bull elephant, with the musket and lance, the club answering for the clapmatches and yearlings, unless where from being frequently disturbed they have become shy, in which case it is necessary to shoot them also. They, too, herd together in rookeries, and between these and the water, from which they are by this means cut off, the hunters advance in Indian file, and when surrounded, turn them up, and drive them to some convenient spot, where, huddled together, they are killed by hundreds at a time with the club. The favorite landings for this animal are the smooth bays with pebble or sandy beaches, but at times they are content with the flat rock. They subsist on birds, fish, and marine productions; they have also a coat of blubber under the skin, which is the toughest and closest fabric of all these amphibious animals, though not one half the thickness of the elephant's in proportion to its size; to cure the skin in a merchantable manner, it should be flinched off with half an inch of fat adhering to it, and after being cleanly washed, and while thus wet, to be spread plentifully with salt, rubbed well into the fat, particularly around the neck and edges, then packed in tiers, or booked up in folds.

Fur Seals. Of these there are three different grades: full aged males, called wigs; the females, clapmatches; those not quite so old, bulls; all the half-grown of both, yearlings; the young of near a year old, called gray or silvered pups, and before their coats are changed to this color, called black pups. This animal is much more sprightly and active on shore than the elephant, leopard, or hair seal; their chief delight is in the heavy surf on a rough or rocky shore; still, by the accompanying plate it will be noticed, they can manage to get themselves quite elevated in the world at times. Here, on the top of a rock forming the N. E. head of Beauchene Island, situated twenty leagues S. E. of the Falkland Islands: they can be seen as having attained a place of apparently perfect security from the attack of those seamen who have landed from the boat at its foot, as the elevation of this rookery is between two and three hundred feet, the ascent thereto being over a succession of shelving rocks, and black and white cliffs. The fur seals and their pups, it will be observed, literally cover the top of this rock; some can also be seen in the water swimming for the shore, some upon the lowest part of the rock, others again half way up, or resting upon the cliffs. A few feet beyond the tussuck grass was another flat rock, covered in a similar manner with seals.

The taking of the animals at this place was commenced at dawn of day; after overcoming the difficulty in climbing up, the seamen were obliged to be very cautious of their footing, the extreme slipperiness rendering them liable every moment to slide off, or else the set on of some bristling old wig (bull seal) while defending his charge, would surely send them there from, down the precipice to instant death. The wigs have long coarse hair, with stiff thin dull colored fur, and frequent mostly the low latitudes on the coasts of Africa and Peru; the second kind are of the middling size, their hair is not so long as the wigs, the fur though is much thicker, and of a finer cast, but reddish hue; they are met with at the Falkland Islands, the islands about Cape Horn, New Holland, New Zealand, the coasts of Patagonia and Chili. The smaller species are a neater animal, not so savage, nor having so sharp pointed a nose as the others; the hair is sleeker and shorter, and the fur, of a rich bright color, is long, thick, and glossy. They are met with at New South Shetlands, west coast of Terra del Fuego, South Georgia, Prince Edward's and Crozett's Islands. All these three different species, strange as it may be thought, have been taken at the Gallipagos Islands, situated on the equator. Over his several mates, as they lay closely huddled around him, the wig keeps a sharp lookout; severe battles frequently take place between the males whenever they approach one another's company; on the other hand, when the females venture to move to another place, or take to the water against the will of the wigs, they are immediately pursued, and by being bitten and shaken, driven back to the starting point; the females have been seen to get some rods off from the shore before their absence was discovered, which was no sooner done than plunging through the surf to all appearance in a great rage, the male has headed them off, biting and driving them back again to where the remainder of his seraglio were quietly looking on without daring to stir.

The clapmatches seldom have more than one young at a time, although sometimes two; it is at this season particularly, that the wigs are very savage, never hesitating to fly at and attack with great spirit, any person who ventures to approach them. They live upon fish, and marine productions; stones also have been found in their maws, as well as in those of the other described animals. They migrate, and with the season return to the shore and herd in rookeries on the rocks, and in the gullies, returning to the water again when the season is over; at this time the animal is very lean, so much so that the skin is become very loose about it; nothing more after this is seen of them until the following season, when they are to be observed coming up again to the shore, exceedingly plump and well filled; where they retire to, to get so fat is something I never could understand; it is also true that they have been met at sea shortly before going on shore in large shoals swimming through the water towards their haunts much like a shoal of herring hogs or porpoises. In calm weather and a smooth sea, they have been seen floating along, hundreds together, and asleep, with but the nose and two of the inflippers, sticking up out of the water, which at a distance appears like the trunk of a tree with its roots, afloat; when caught thus asleep, they can easily be taken by the harpoon or spear, by approaching them silently.

These fur seals are the animals the celebrated Captain Cook and other ancient voyagers call sea bears, sea wolves, &c. While waiting at the resorting places of these animals, they have been seen far off on the sea as the eye could reach, coming directly and quickly towards the land, springing out of the water and then plunging in again, in a regular undulating movement, and after passing through the surf, land on the rocks, where being arrived, a few moments are passed in looking about for a suitable spot, the body then moving off to their places in the rookery, a mode of proceeding very similar to that of the hair seals. The wigs, generally come up in smaller flocks a few days before the clapmatches, and as these last arrive, select their mates from among and drive them off to their location directly.

The skins are taken in the same way as those of the sea lions are. If these seals are to be skinned for the China market, the fat, and portions of the meat should be taken off with the skin, the whole then washed and spread on the rocks to drean, afterwards taken to the beamer, where the fat and blubber is cleanly beamed off, the flipper holes sewed up with twine, and the skin stretched out by means of ten wooden pins (which to be of a suitable size, should be half an inch in diameter at the head, and twelve or fifteen inches in length) on the turf to dry, and when perfectly dried are in order for that market. But for the European, London, or American markets, the skin should be flinched off, taking half to one inch thickness of the fat with it, then washed perfectly clean from the blood and dirt, and while wet plentifully salted, and booked or kentched away. The young fur seals, after they are six weeks old, and until yearlings, are used as, and considered to be, good fresh meat for roasting, frying, broiling, or stewing; in fact, persons have preferred them to lamb. The flippers, when well dressed and cooked, are not inferior to the callipee of the best green turtle.

CHAPTER XIX

VOYAGE OF THE SHIP "VOLUNTEER"

ON the 5th day of June, 1815, the ship *Volunteer*, well armed, and fitted with a company of thirty, officers, men, and apprentice boys, under the author's command, left Sandy Hook, bound on a voyage to the South Seas and Pacific, after fur seals, sandal wood, &c. for the Canton market. Arrived at Port Lewis, at the Falkland Islands, a small vessel rising thirty tons was built, the frame and materials for which had been taken out in the ship, and an officer with eight men left in charge of her, to take seals during the absence of the ship to the Pacific after sandal wood.

As a caution to strangers on opening and sailing, or working up Port Lewis Bay, I would remark, that about midway up, a cluster of tussuc islands, called the sea-lion islands, will be noticed, the passage on either side of them appearing to the eye to be fair and clear of danger, and that on the south of the islands, the most preferable; this, however, must be avoided, as in its so being we were most sadly deceived, for as the *Volunteer*, with a fine whole sail breeze from the northward, passed around the northern promontory, she then opened and steered up the bay; on nearing, and getting a fair view of the above tussuc islands, the passage on the south appeared to be the most eligible; we, therefore, took it, and when arrived abreast of the east end of the largest island, furled the staysails, with the fore and mizen top-gallant sails; the wind, though moderate, a leading breeze, with smooth water, and the ship going at the rate of four or five miles an hour; one officer was on the lookout at the fore-top gallant yard, another at the jib boom end, and a hand in each side channel at the lead, so that when one lead was up, the other should be down. Notwithstanding all this precaution, when the ship had passed up a little above the middle of the island, the water, from twelve fathoms muddy bottom, began to shoalen at each cast; still the lookouts could not discover any danger; but when the seaman at the larboard lead announced "and a half four!" which was directly echoed by the other, and sung "by the mark three!" the helm was ordered a lee, the ship came promptly to, and as her sails began to shiver, took the ground, and was fast ashore on a sand bar; all the sails were instantly hove aback, and preparations made to carry out an anchor a stern; as the tide, however, was now on the rise, before this could be effected, the force of the wind on the sails backed the ship off afloat, and as she gained the depth of ten fathoms came to anchor, with mud and clay bottom. We then furled the sails and sent a boat to sound, when it was found that a hard sand bar extended from the island entirely across the passage to a point on the main, having then, at about half tide, on its ridge, only three feet water; here the ship remained at anchor the rest of the day and through the night, while in the meantime an officer in the boat had sounded out the north passage, which being found free from danger, we weighed anchor, and returned again around the east end of the larger island, and under single reefed topsails, with a gale from the W. N. W. worked the ship up the passage on the north side, and came to, at the port. The bottom, as far as our knowledge extended, generally throughout this bay, is good anchoring ground, keeping clear in all the different depths of water of seaweed or kelp; we discovered no shoal or shallow water, but what the kelp designated, except the above mentioned bar. To the left of the before mentioned false south passage, as you advance up the bay, and abreast of a beach, is the spot where the French corvette, *L'Uranie*, Commodore Freycenet, was run on shore and wrecked, being at the time bound on his astronomical voyage, &c., round the world.

During the three months while our crew were employed at these islands in building the before mentioned vessel, and in taking seals, our hunters succeeded in procuring for the ship's company, two hundred and ninety-two wild hogs and pigs, nine hundred and thirty-seven wild geese, some ducks, teals, also seventy-three barrels of albatrosses' and penguins' eggs, together with about five barrels of excellent mullet fish, from twenty to twenty-five inches in length, and some shell fish, which were very easily obtained at most of the harbors. This wild pork was salted and barreled up, and was preferred by the crew to the salt pork laid in at New York.

December 10th, 1815. Got the ship under way from the Falkland Islands, and proceeded round Cape Horn into the Pacific. After searching for and taking some fur seal skins at the Island of Massafuero, her course was shaped upon a wind for the coast of Chili, with a view to put into the port of Valparaiso, to refit ship and obtain water and refreshments, but the wind not enabling the ship to fetch that port, it was thought best to bear up for Coquimbo. While standing in with the coast for this purpose, the weather so thick and cloudy as to prevent our taking an observation, by which the correct latitude might have been ascertained, the shore, or rather firstly the surf on it, was made. After doubling a rocky islet, which lay off a projecting point of the land, we stood into a bay, which was supposed to be that of Coquimbo, and anchored in twelve fathoms water, near to a beach at the bottom of the bay; this was near sunsetting, the weather still thick and smoky, and preventing our seeing far around; nor was our mistake discovered, until we noticed there was no appearance of habitations on the shore, on which some persons were seen, when for the first it became doubtful whether this was the bay we supposed it to be; this point was settled upon proceeding in the boat to the shore, by the information of those persons there met with, from whom the ship's true situation was ascertained to be in a bay very similar to that of Coquimbo, a few miles south of it; night now closing, we remained at anchor here until the following morning. Being put upon my guard, from an acquaintance previously acquired, of the jealous disposition of this people, and a remembrance of the detention and difficulty on my former stopping at Valparaiso, all communication with the shore was prohibited, the inquiry above having been made without landing from the boat, or any one of these people coming off to her.

Next morning we proceeded round the point to the north, into the bay of Coquimbo, and came to an anchor in the harbor near to, and abreast of the fort and battery. Being unacquainted with the recent revolution in Chili, and change of government into the hands of the royal party, we were much surprised at seeing a number of troops on shore in the royal Spanish uniform; but still as the purser, the author's only son, both spoke and wrote the Spanish language equal to a Castilian, the boatswain also speaking it, sufficiently well for ordinary purposes, we felt confident that no misunderstanding could arise between the inhabitants and ourselves. Having in charge letters for the United States' Consul General. Matthew A. Havel, at St. Jago, the capital of Chili, I had prepared the following letter to this gentleman, on business relating to the ship, and inclosed the others in it.

COQUIMBO, FEB. 8TH, 1816.

Matthew A. Havel, Esq., United States Consul General at St. Jago, Chili.

SIR,—Having put into this port, for the purpose of overhauling my ship, filling water, and obtaining refreshments, I have herewith the honor to inclose your letters, intrusted to my care, together with the latest American newspapers, and also to tender to yourself the first offer and preference, for any freight my ship may be able to take; therefore, should you alone, or in connection with any Spanish merchants, wish to freight the *Volunteer* from Valparaiso to New York, or to any port in Europe, to Canton, and on to New York, or to return from Canton to Valparaiso, or any other port in Chili, please state what such freights would consist of, and on what conditions they could be engaged, should I be at Valparaiso in July or August next, to accept of either of them; being now bound to the leeward seal islands to take seals. Be pleased also to give an immediate answer by letter to the above propositions, that I may govern myself accordingly; and inform me if those freights can be taken in the *Volunteer,* she being an American bottom, consistent with the Spanish commercial laws and regulations of Chili, as no reward or inducement would engage me in any illegal trade: the *Volunteer* is a new and first rate ship of her class, coppered and armed. If agreeable, please mention your friend or agent at Valparaiso, that I may wait on him on my arrival there. Inclosed is likewise a copy of my instructions from J. Byers, Esq., agent for the owners of my ship; an immediate attention and answer to the foregoing, as I shall wait here a few days, will much oblige.

Very Respectfully,
Your Obt. Servant,
E. FANNING,

Captain of the American ship "Volunteer."

At the same time, and by the same conveyance, the following letter was addressed to the governor of Coquimbo.

On board the American Ship "Volunteer,"
at Coquimbo, Feb. 8th, 1816.

To His Excellency, the Governor of Coquimbo.

SIR,—I take the liberty of intruding upon the goodness of your Excellency, so much, as to request your aid in forwarding to St. Jago, by the first opportunity, the inclosed packet, directed to Don Matthew A. Havel, American Consul General at that city, and also the additional favor, that your Excellency will be pleased to grant me permission to remain at anchor in Coquimbo, to overhaul, caulk, and refit my ship for sea again; also to procure water and refreshments, and remain until I shall receive an acknowledgment of the receipt of the packet for Don Havel. Your Excellency's compliance will greatly oblige.

With much respect,
Your Obt. Servant,
E. FANNING,

Captain of the American ship "Volunteer."

Accordingly so soon as the ship was brought to an anchor, manned the boat and went on shore, accompanied by the purser. On having the commandant pointed out to me, I requested permission from him to send an officer to the city, which is in sight, and situated a few miles from the harbor; the road leading to it being round the head of the bay, and nearly level; so that travellers can be seen from the ships in the harbor, a good part of the way on it: this request, with another for a horse, was granted, and handing to the purser the packet, he proceeded for the city, to deliver it to the governor and return with an answer with all despatch. It was 3 P. M. when the purser started off, he having a Spanish officer riding on each side, with two mounted soldiers a few rods in front, and two in the rear; this, as the commandant, through his interpreter, informed me, was as a guard of honor, because the business was with the governor; the distance to the city being only about one hour's ride, of course I could reasonably expect the return of the purser before sunset.

Joseph Maria Gomez (for such he gave me to know was his name) then stated he was the commandant of the fort, and had command of that division of the royal Spanish army stationed here, and that by the recent revolution, which had taken place, Chili was, thanks be to God and the saints, said he, once more brought under subjection to his master, King Ferdinand VII. On my mentioning the cause which had brought us into their port, and the particulars of what I had written to the governor, he said all would be readily granted, and that I was at liberty to commence overhauling and dismantling my ship as soon as I pleased. On this I took leave and returned on board, being anxious to have all ready for sea upon the receipt of the Consul's answer, which it was said would probably be in ten or fifteen days. We therefore unbent the sails, and commenced sending down on deck the spars and rigging, and overhauling them.

At 9 P. M. finding the purser did not return, sent an officer on shore to make inquiries after him; who, upon his return, reported he had not been able to learn why the purser had not yet returned, or any thing about him, but through the boatswain he had found out, the return of those two officers who had started off with the purser for the city, as well as that General J. Maria Gomez had proceeded up to town; in addition to this, that two heavy pieces (forty-two pounders) of cannon had been brought to the fort, and were then being placed in it, in addition to those of a smaller calibre already there. This information, with the fact that sentries were placed on every prominent spot around the harbor, caused me to suspect all was not right, and raised a doubt in my mind that the concessions already obtained, were not made with a friendly intent; as the case then stood, I concluded to wait with patience for what should transpire in the morning. The next day before I had risen, the officer in command of the deck, reported the marching of more troops, then in sight on the road from the city, to the harbor, and soon after rising from breakfast, the following note was handed to me, accompanied by a verbal invitation, couched in very polite language, from the bearer, a military officer, that the commandant Gomez was waiting to have an interview on shore. The note was as follows:

Coquimbo, Feb. 9th, 1816,
2 o'clock, A. M.

To Captain Edmund Fanning, of the ship "Volunteer."

Sir,—The conductor of this comes commissioned from me, and from this city, to answer you respecting the points contained in yours of yesterday, which I inform you of, and also the name of the gentleman, Don Jose Maria Gomez.

Yours, &c.,

Juan Antonio Olate.

Upon this I ordered the boat to be manned, and in her, accompanied by an officer and the boatswain as interpreter, proceeded to the shore, still thinking it very strange that the purser had not returned. On the shore, J. Maria Gomez, attended by a few officers, was awaiting; while in different places small bodies of troops were seen paraded. On stepping ashore, and advancing a few paces towards J. Maria Gomez, one of these detached bodies of soldiers instantly surrounded me with charged bayonets, and marched me off to prison, another party seizing upon the officer and boatswain, and taking them also to prison, and the men to a third place of confinement.

After remaining some time alone in my solitary cell, in the midst of swarms of vermin, not altogether in the best humor imaginable, my meditations were disturbed, by the entrance of a fat padre and his burning tapers, with some official character or other, and an interpreter. *Cappetain Amerricana,* said the padre, *usted confesso; Frigat Amerricana, mo, yeo, mahlah.* (Meaning, I come for you to confess that your intentions with the American ship are bad). Such he flattered himself was very good English; but notwithstanding all this, I thought proper not to answer, and therefore remained silent, my looks giving them to understand their visit was not considered either friendly or welcome. The ghostly father was evidently not pleased with the reception, and speedily all three took their leave. A little while after, an officer entered, and stated that he had orders to escort me to the commander's quarters; my double guard, as we left the prison, being joined by those of the escort, the whole were formed into a hollow square, myself in the centre, and in this manner conducted to headquarters, on the way passing a number of squadrons of horse and foot, under arms, and paraded to show off, or some other purpose. Arrived at our destination, I was greatly surprised to see my son, the purser, for the first time since he had been sent to the city, standing on the piazza, with a few officers; seeing me approach, he turned and made some observation to one of those officers, who then spoke in Spanish to the leader of the escort, and he in turn giving an order, by which the escort altered from the square to two lines in single file, and then halted. Having permission to speak to me, my son approached, and as I took his hand, whispered, "Father! I have overheard the Spanish officers' intentions; they are going to execute you;" and then burst into a flood of tears.

Although greatly shocked at this unexpected piece of information, I was still enabled to retain sufficient command of myself, to request him also to show to our enemies we have American hearts, and fortitude enough to bear up against their injustice. He instantly perceived the propriety of this, and becoming composed, stated, that the moment he had arrived at the city gate, the officers who were with him had put their horses at full speed, he also doing so; in this way riding through the streets, they came to the residence of the governor, where he was then ushered into the presence of his Excellency, in whose company were several military officers and other gentlemen. On receiving the packet, the governor broke the seal, and perceiving the inclosed despatch for the United States' Consul General, immediately thereupon ordered an officer in attendance to take him in charge, and imprison him forthwith, being thereafter kept in close confinement, until taken out to be carried to the harbor, in compliance with the author's demand. Since this he had, while in conversation with several of the officers, some of whom were sociable and communicative, learned that they entertained strong suspicions, and indeed were firmly convinced, we were engaged with the ships in assisting the cause of their opponents, the Patriots, supposing when she anchored in the south bay, and remained there during the night, it was on no other business than this. He farther mentioned, that the governor had early this morning held a council, at which, by what he had accidentally overheard, some of the officers said my immediate execution had been determined upon; by this promptness of action, they imagined the officers and crew would be terrified, and at once confess their designs and engagements with the Patriots.

To this I replied, admitting all they have determined upon should be carried into effect, still there was much satisfaction in the knowledge of our entire innocence; but to suffer ourselves to appear deficient in firmness, would not only bring a blemish to the American character, without assisting us at all, but also would be the very worst policy to take to relieve ourselves from the present disagreeable and painful situation; for should these blood-thirsty barbarians, as from mere jealousy and suspicion they had determined to do, actually take our lives, a departure from truth, or timidity and fear, could not save us; but let us meet the worst, and leave our cause to God and our country; then taking him by the hand, I requested him to say to the officers and crew of our ship, my wish was that they also would remain firm to truth and their innocence, nor suffer any power on earth to sway them from such determination. As we parted, I had the satisfaction of seeing that my remarks were not lost, for the countenance of the purser was calm and collected, and his whole appearance that of a man ready to meet whatever evil his enemies might put before him.

I was then conducted into the hall at headquarters, each side of which was occupied by double rows of soldiers, while at the farther end was seated the commander, Gomez, two officers, and an aged gentleman, of a benign countenance, in citizen's dress (this person was pointed out to me as the King's Judge Advocate); a semi-circle of military officers ranged on each side of the judge, and a person acting as scribe seated at a table in front of these, completing the company. As soon as they seemed to be prepared, the chief, Gomez, made some remark in Spanish to his interpreter (an officer, judging from his speech, born in Ireland), this last then turned, and in English gave me to know, this was none other, than the King's court, which was now going to interrogate me upon the nefarious employment myself and ship had been engaged in. This insulting observation had well nigh produced a retort, but the better policy now appeared to be, to remain silent, and enter my protest.

I accordingly addressed the chief Gomez, with a look and voice intended to convince him of my sincerity, that notwithstanding the many armed gentlemen around, and the glittering bayonets he chose to have present, I protested, in the name of my government, against all these acts and proceedings; and farther, that I did not conceive myself bound, and would not answer any interrogations, other than when my interpreter should be present. This was indispensable to our correctly understanding each other, and the only fair way of proceeding.

Don Ignacio Borgues, whose conduct has subsequently proved him, what his countenance bespoke him to be, a man possessed of a feeling and kind disposition, and an upright judge, beloved and respected by his countrymen, and maintaining through all the revolutions and changes in this country, a high station in point of character and talents, was now the King's Judge Advocate. His seat, or rather palace, with its beautiful garden in which were a variety of flowers, and the richest fruits, was situated on the road leading from the harbor to the city. Immediately after the breaking up of the above mentioned court, our purser received an invitation to visit the old gentleman, which he accepted, being at this, and every succeeding visit, treated with the utmost politeness and kindness by the family.

An animated conversation now took place between the commandant and the King's Judge Advocate, which last asserted the captain to be in the right, when insisting upon having his own interpreter present at the interrogations, as well as being entitled to counsel, if he desired it. These were very unwelcome facts for the chief, Gomez to learn, and instantly becoming much enraged, he dismissed the court without any more ado. The officer in charge of the escort then received his orders, and marched me off to a place by the parade-ground of the troops; here, after going through the fuss, as it proved to be, of a sham execution, I was remanded back to prison. Shortly after, Gomez, with an interpreter, and a soldier on each side, with musket, and bayonet charged, entered my prison, and asked if I wished any thing to eat or drink. My mood was not very pleasant at the time, so that he received a very short answer in the negative; he then inquired what I did wish for, all the while standing at as great distance as the walls of my small cell would permit, as though I was a tiger, or some other wild beast, evidently not trusting much to his guards. I then requested the use of pen, ink, and paper; this was granted, and with these I at once addressed a note to the governor, reciting what had taken place and demanding if Spain and the United States were at war that myself, officers and crew be treated as prisoners of war and not as the worst of criminals, at the same time assuring his Excellency that our sufferings and the indignity shown us should be laid before our government.

The bearer of my note soon returned with a message from the governor that if I preferred remaining a prisoner on board my ship, full liberty was now given me to repair on board. This I declined to do unless my officers and men went with me and this decided stand soon brought a general permission for us all to return on board.

Arrived once more on board the ship, we ascertained that all hands had been taken on shore and imprisoned, except the chief officer, the steward, and a boy, a company of soldiers being placed in charge of the vessel; by these, every thing was displaced and tumbled about; in the hold, the casks of stores, such as the molasses, the liquor casks, &c., were bored by gimblets, and the contents suffered to run to waste on the decks and about, until the steward managed to stop them; in this spirit of maliciousness prosecuting their search and examination, for some evidence to prove us connected with the Patriot's cause, the remonstrance and entreaties of our officer being of no avail.

The sabbath following our imprisonment, the American colors were set upon the ship, as usual, out of respect for the day, the men also being at leisure; this was noticed by General Gomez, who wished to know, inasmuch as the company were thus idle, whether there could be any objection to their all being examined on shore, merely as a matter of form, and as his excuse to the Vice-King, he was informed that nothing of the kind would be voluntarily consented to, nor in any case should an examination be admitted, unless my interpreter, the purser, was returned to us from the city, and present during the same; this being done, that is the purser first liberated and sent on board, the proposition would be further considered. Accordingly, in the course of the forenoon, the purser was set free, and returned to the ship. Early in the afternoon a launch pulled up alongside the ship, filled with troops; these soon scrambled up over the rail, and were formed in double line on the deck. After they had all primed and loaded their pieces, and fixed bayonets, a detachment was ordered to the quarter-deck, where I was quietly walking, all the while noticing their strange and unaccountable conduct. Being surrounded by this squad, their commandant drawing his sword, advanced to me, saying, "Sir, you are my prisoner!" "Then, sir," I replied, "I surrender my ship;" and accordingly gave my first officer charge to haul down the colors. This being done, caused the Spaniard to hesitate, and Gomez, an attentive observer of all that was passing, who at the moment was standing on the bank within pistol shot distance, called out to know why the colors were thus hauled down, receiving as a reply, the captain has surrendered the ship.

Except our second officer and the steward, all the ship's company were ordered into the launch, part of the troops getting in also, the remainder staying on board. On landing from this launch with my son, the purser, by my side, we commenced our walk on to headquarters, and arrived in a few minutes after at the hall, a place about forty feet by twenty. On entering it, we found our first officer, Mr. B. Pendleton, seated uncovered, on one side, while Jose Maria Gomez, in the style of a braggadocia, was marching backwards and forwards in front of him, brandishing his hanger about very valiantly, each time as he passed Pendleton, cutting within an inch or two of his head. I know not whether such conduct was most calculated to excite feelings of contempt or anger; as it was, the first in a very strong degree were predominant, and the man's noise and bustling about, were it not for the serious transactions in which he was concerned, would have called forth a hearty laugh. This strutting was undoubtedly intended to produce fear and terror in Mr. Pendleton's mind, as well as his fellow-sufferers; it was labor in vain, however, and that he (Gomez) might know the estimation in which we held such conduct, the purser was directed to tell him we conceived his behavior not only unbecoming a gentleman, but also disgraceful to the commission which he held in the service of King Ferdinand VII, his master. The effect was as desired, for being delivered in a clear and audible tone, and in the Spanish language, all the troops, and others in the hall, must have heard the words. Gomez instantly came to a stop, and colored both red and white at being thus exposed. I then farther told him it was our expectation, if we were not considered prisoners, to be without loss of time returned to the vessel, and then left the hall in company with the purser, but stopped at a house near by, where some Spanish officers of his acquaintance were collected together; in a few moments the officers and crew were seen repairing to the shore, and embarking for the ship, being also allowed to depart by those in power, the soldiers who had had charge of the ship then returning to the shore in the launch; still, however, a man and boy it was discovered were on shore, and for these, persons were despatched, who succeeded in discovering them at the commander's quarters, where they were being examined, but on the remonstrance of the purser they were given up. It then appeared that this man, while leaving with the others, had been induced to stay, and after being well plied with wine and fruit, had been taken before the chief, Gomez, for the purpose of being interrogated in relation to the ship's voyage, of what had been done, and what was going to be done; this history he had nearly finished when the purser found him.

The day after these occurrences had transpired, an order was brought from the governor requiring the unhanging of the *Volunteer's* rudder, the same to be sent on shore, while at the same time a guard of soldiers were to be sent on board ship. The last body of these fellows had sufficiently incommoded us, so that I determined not to acquiesce in this demand and at once sent a note to the governor protesting against his order and demanding an explanation of what had taken place.

Gomez, the commandant, had at this time prohibited every article, even the most trifling kind of fruit or vegetables, from going on board our ship; this prohibition was rendered doubly harsh, from the fact that our crew were laboring under a scorbutic affection, which the fruit would have removed. Notwithstanding these orders, when young Borgues, son of the king's judge advocate, came on board with the bearer of the governor's last order, he brought with him, as a present (and one highly acceptable) from his parent, two large trunks, one filled with delicious peaches, grapes, &c., and the other with salad, onions, and other vegetables. Gomez, with a threat, had forbid these being taken on board, but the young gentleman producing his father's written instructions, informed the chief he would deliver them according to their destination at all hazards, holding himself responsible for the consequences.

The next morning I received an evasive communication from the governor but at the same time came an order to supply the ship daily with fresh meat, fruit and vegetables as I might direct. After a few days things began to wear a smoother front; our supplies of refreshments, although in a somewhat limited quantity, were getting on board; the refitting business too, was progressing, all looked well until the afternoon of the 16th, when the second edition of the same troubles were visited upon us by an officer and a party of soldiers, who pulled up alongside, with a demand that all our powder should be delivered up to him, adding, if the same was not peacefully surrendered, his orders were to take it by force. He received as a reply, that no consent on our part could be voluntarily granted, for so long as the arms and sails of the vessel were retained in their possession on shore she must be considered as at their risk, and the crew as prisoners of war. The soldiers then mounted to, and were paraded on the deck, and ordered to load their pieces; after which, they were formed from the gangway around the quarter deck, and encircling the companionway, they then commenced passing the powder up, from below, thence along through this line of troops to the gangway into the launch. A short stop was put to this disposition of our property by the timidity of the lieutenant, or the person who held the second command of the troops; his fears were occasioned by the sparks which flew from the armorer's forge, who was at this time busy about the same, upon some necessary work, having been unable to obtain permission to erect it on shore, it had been put up on the main deck about midships; sparks flying, and powder passing, at last induced the officer to call out to his senior and notify him of his unwillingness further to risk his own and the lives of his men until the forge was stopped; the senior at this hint wished the request might be granted, but was told no assistance on our part could be rendered him in the fulfillment of his orders. "But," he replied, somewhat astonished at our indifference, "you will all be blown to heaven then." This, we answered, would be all very well, as we should also, part of the way at least, have their company. Muttering something about North Americans, strangers to fear, more so than any men on earth, captain and all indifferent or unconcerned about being blown up, &c., he turned round to the purser and begged his intercession; this was granted by his advising him to order a file of his men to charge bayonet upon the armorer, who very probably would retreat from his forge, and they could then put the fire out. "He does not understand your language," said the purser, "so that you can bid the troops not to harm him in their charge, without his knowing of the order." After literally following this advice, the armorer of course retreating, and they quenching the fire; the officer thanked the purser for it, and then, completing the disembarking of the powder, left the ship with all his troops for the shore. A demand to the governor to ascertain the meaning of this fresh indignity brought no reply and all remained unexplained as far as up to the 22d; meanwhile our ship was completely prepared for sea, with a stock of water on board, wanting only her sails, powder, and arms, to proceed with.

I now hit upon the expedient of writing a particular account of all the facts of my detention and sending it to the governor with the request that it be forwarded by way of the isthmus of Panama to the United States government at Washington. With this statement was sent a translation into the Spanish language and both were left unsealed.

The translated copies, as the purser was informed,

the governor immediately by advice of his courier, sent off, by an extraordinary courier, to the Vice-King at St. Jago, the ship meanwhile to be detained for an answer, which could not be short of seven to ten days. The council also, at this setting, decided forthwith to take off all restrictions from the ship, giving the officers and men permission to come on shore, go where, and return when, they liked; myself also receiving liberty to procure from the shore what I pleased. The governor's orders, in accordance with these, was brought by the officer, who returned with the purser; Gomez hence, in his conduct and bearing towards us, was entirely altered, but knowing somewhat of his character, these professions of friendship were duly estimated, and he credited accordingly.

On March 8th, I received a communication from the governor stating that the supreme government had directed him to deliver up everything that had been removed from my ship and that I was at liberty to leave the port.

Our sails, powder, and arms were then returned, and a supply of refreshments, as wanted, was furnished without charge; thus all ready for sea, I notified the commandant, J. Maria Gomez, in the evening, that the ship would proceed with the first fair wind.

Early the next morning, an officer from the city called on board, with the compliments of the governor, saying, that agreeable to the orders he had received from the supreme governor at St. Jago, he would have the honor to make a visit on board the ship in the afternoon, to apologise in person for the reception and treatment we had met with. This was quite unexpected; and accordingly at 4 P. M. his Excellency, attended by a numerous retinue of military and civil officers, came on board, and with every appearance of friendliness made the promised apology; he farther mentioned his hope, that upon reflection, I should conclude not to represent the affair to our government, as he could say for himself, that he had been overswayed by his council, and were it to become a national affair, he alone would suffer.

It afterwards appeared, that the packet sent on the first day, directed to Mr. Havel, the Consul, had instantly excited their suspicions, that it was in some wise calculated to act against the Royal cause, and advance that of the Patriots, and was thus the means of bringing down upon us their unwarrantable and severe treatment; but when this, with the letter inclosing it, had been forwarded to St. Jago, and examined by the supreme governor there residing, their error was discovered; and hence the anxiety of the government, in directing their apology through the governor of Coquimbo, in this way endeavoring to remove, as far as was in their power, the insult and wrong committed by their premature and disgraceful acts, especially as they now began seriously to fear our government would likely take the matter in hand. After these ceremonies, apologies, and leave taking were over, we got under way from Coquimbo, and proceeded on our voyage, standing to the northward. After visiting the Lobos Islands, we repaired to the Gallipagos, so much celebrated from being the resorting place for the early buccaneers.

Here we obtained about 8,000 fur, and 2,000 hair

seal skins, and also a good sea stock of the celebrated land terrapin (named by Commodore Porter, the elephant terrapin), and several bushels of delicious pears, gathered from the prickly pear tree, which grows in abundance on several of these islands; it is a size less than the New York bergaloo, but of a similar shape, with one side a bright golden color, and the other a bright crimson; it is an effective anti-scorbutic. This excellent fruit we gathered in a very careful manner, to avoid bruising it, by hooking it from the trees with our small iron seal hooks, fastened to a pole ten feet in length, and as it fell from the trees then received into a sheet, held at the corners by four seamen. After thus procuring them, they were placed in our nettings, between the cabin carlings or beams, and kept good for over two months. When ripe and mellow, this pear is equal in richness and palatableness to the best pear or peach from the New York market; when cooked in pies or tarts, according to our notion, they were equally nice, and relished as well as any to be procured at a confectioner's. After receiving these on board, we took our departure from the Gallipagos, and steering to the south, arrived at the Island of St. Mary's on the coast of Chili, where we came to anchor, and took on board about 14,000 more seal skins.

While the ship lay here, our poor steward received a severe scald from a pot of boiling hot tea, which one of the men had just taken out of the galley kettle and was carrying below; this was at the dusk of evening, the steward not perceiving the seaman's approach, was coming up the gangway ladder, and striking his head against the pot turned it over, and being uncovered, the scalding contents were thrown on his head, down the neck, back, and shoulders, making so severe a burn that the back part of the scalp, with the skin, came off in the attempt to dress it, leaving a raw and deep wound. The boat very fortunately had just before this misfortune, brought on board a quantity of greens which had been gathered on the island; among the lot was a parcel of very large leaves of the dock herb, very similar to that growing in our fields about the barns; having heard that this was a healing herb, I determined to apply the same to the steward's wound, which was forthwith done, after first causing a strong decoction to be made by pouring boiling water on the green leaves, and simmering them a short time over a slow fire, then, after this preparation had cooled, washing the wound with it; after bruising the fibres of the same leaves to make them soft, and dipping them in the decoction, they were spread over the wound two or three thicknesses. It was astonishing how soon this application healed the wound, for not only was immediate relief from the pain and smarting obtained, but a new skin began to grow over without leaving scarcely the appearance of a scar; what added to the surprise was, that the hair grew on immediately the same as before: the dressings and washings were repeated every few hours, or as often as the leaves lost their moisture and became somewhat dry.

Another accident of this kind occurred at New Island, one of the western Falklands, where the ship was moored while waiting for the completion of her cargo. At this time a party of men under command of the first officer, Mr. B. Pendleton, were out on a sealing excursion, and had at evening pitched their encampment at a low island some twenty miles from the ship; the surface of this island was covered with coarse grass growing in the tussuc bogs which are of various sizes, and spread near each other; the top of this grass is some feet above man's height, and the ground between the bogs covered with a thick layer of the dead grass, the growth of previous years, and so combustible withal, as when once on fire there is no possibility of extinguishing it. On such excursions as this, the crew after hauling up and securing their boats, fall to work to make a sleeping hut (or nest, as they call it); for this purpose two large bogs are chosen at a good distance apart, and the inner borders of the grass then cut down fairly so as it were to form the walls; after this, the tops of the grass on the inner margin of each bog is strongly tyed together to form the room; over this is placed a thatch of grass of sufficient thickness to keep out the rain, be it ever so violent; each end of the hut is then walled up tightly with portions of the bog, leaving at one end but a small opening by which to enter the premises; this again, when the crew are in, is closed by means of a dried skin secured to its place with wooden pins, the interior having a layer of the dried grass to answer for beds, so that when in, the men sleep warm and comfortably, and at times are rather loath to come out. These huts are placed as near each other as possible, or as the bogs will allow; while they are building, the cook makes his fire upon the beach, and prepares a supper of meat, with a cup of tea for each, all turning in after partaking of the same, except himself, who remains by the fire to extinguish it, and prepare breakfast in the morning. This our cook, a good natured careful body, had always done; it appeared, however, that after extinguishing his fire, he was in the habit, unknown to the officers, of enjoying a comfortable smoke of his pipe after retiring with his mess to their hut. He had so done on this occasion, and after supposing his pipe was out, had laid it at his head as usual, but had not got sound asleep before it was discovered that the grass was on fire, no doubt originating from a spark out of his pipe; unable to smother this, he awoke his shipmates to his assistance; their efforts failing, the officers were then called, who instantly alarmed the whole encampment. From one of the rear huts a young man (B. S. Cutler) was one of the last to awake, by which time the flames had nearly reached them, and as he, came out of the hut, not yet fairly awake, he became bewildered, and seeing the fire spreading around in front, retreated towards the interior of the island, but recovering his recollection he mounted to the top of one of the tussuc bogs to take a survey, whence observing himself nearly surrounded by the flames, he concluded his shortest and safest way was to retrace his steps, and if possible, pass through the sheet of fire to the boats; this he proceeded to effect but having over his other garments a frock highly charged with oil from the fat or blubber of the seal skins which he had worked in, he had not more than entered the flames before it took fire, so that by the time he had crossed the burning grass, all his clothes were burnt to a cinder, and his body and limbs completely roasted, so much so, that after he had been plunged in the water and taken out, the skin cracked and came off with flesh attached to the remaining portions of the dress; having, however, taken care to hold one of his hands over his mouth and nose while in the flames, he had imbibed but little if any of it.

Mr. Pendleton having heard that oil and fresh skins of animals was a good application in giving relief from burns, directed several of the penguins then near by to be killed, and their skins to be taken off with about half an inch in thickness of the fat and flesh attached; binding them in this state around the roasted body of the young man, an immediate relief from pain resulted from it; then preparing a litter, the young man was forthwith taken to the ship, where finding he had experienced so much benefit from the first application, I directed fresh skins to be brought from the shore, and in this way for ten days we replaced the dry with fresh skins twice in every twenty-four hours, the body the while being kept gently open, and the patient living on a light diet of gruel, soup, &c. No other application was made use of, and the rapidity of his recovery was truly astonishing, for a new skin like as of an infant grew over his person, and in one month's time he could move alone about the deck, and shortly after attend to his duty again. Those skins (and would not those of any fat animal have the same healing effect?) were soft to his wounds, and kept him always free from pain, except at the time of removing and replacing them.

From St. Mary's the ship proceeded to the Falkland Islands, where we made a short stay, to procure some additional seal skins and oil, then finally got under way for New York, at which port she safely arrived, April 13th, 1817, after an absence of twenty-two months. On the ship's arrival, a piece appeared in the newspapers, giving a sketch of the proceedings against the *Volunteer* and her company, as well as of the seizure and detention of other American vessels in ports on the Chilian coast; upon this a messenger sent by President Monroe from the department of state at Washington, sanctioned by the signature of Mr. Rush, acting secretary, waited upon the author, and requested the particulars in relation to these occurrences, with at the same time an opinion as to what measures he could recommend, most likely to prevent their repetition. This was most cheerfully given, the sending a naval force to the coasts of Chili and Peru, being advised in the strongest terms, as the only way, in the present unsettled state of those countries of putting a stop to farther depredations, and preventing our commerce, trade, and fishery, from being destroyed in that quarter. Shortly after this, Mr. Worthington returned from Washington, and wished assistance in getting to the Rio de la Plata, as he was anxious to arrive at Buenos Ayres as quickly as possible, thence to travel over the mountains to Chili, on confidential business for the government, which had its origin out of my communications sent to Washington. Mr. Worthington also stated, President Monroe had expressed himself as being much gratified with the stand which had been taken by the *Volunteer* at Coquimbo, and upon the author's recommendation, had decided upon sending an United States ship of war to the Pacific. For this piece of service, the sloop *Ontario*, Captain Biddle, was chosen, and ordered to be in readiness for sea with all despatch. Thus it was the weighty evidence our case produced to government, and reasons advanced by myself, for making the Pacific a naval station, that opened the gate, as it were, to induce government to place there a naval force; and should this have been any advantage to his fellow-citizens, particularly to the officers of our navy, the author feels confident, that in their known liberality, they will give him due credit. All the American vessels seized in other ports on this coast, as has been subsequently ascertained, were promptly released, through the exertions of that gallant naval officer, Commodore Biddle.

This was the first voyage of the *Volunteer;* on it the ship was found to be a fast sailer, and as safe and agreeable a sea-boat as any mariner need wish for. She was built at Stonington, Connecticut, by that ingenious shipwright and builder, Benjamin Morrell.

CHAPTER XX

SUNDRY VOYAGES TO THE SOUTH SEAS

SEPTEMBER 4th, 1817. The ship *Sea Fox*, under the command of the author, sailed from Sandy Hook for the South Seas, touching first at Byers' Island, a group situated on the E. S. E. coast of the Spanish Malone, or Eastern Great Falkland Island, and so named after James Byers, Esq., resident in Springfield, Mass., an ardent supporter of American enterprise, and principal owner of the ship and her tender, which first examined them; here a cargo of elephant oil and seal skins was procured for the ship.

The annexed plate represents an encampment of sealers at these islands; in the foreground, part of the crew are engaged in preparing a supper of upland geese; some of this game lies at the feet of the two officers, who are leaning against the rocks, engaged in conversation; opposite to them a seaman is picking a goose, while another is dipping some loggerhead ducks in a kettle of boiling water, for it is necessary to scald these birds to enable the men to pick them; others of the crew are backing skins from the landing, a short distance round the point, the situation for rendezvous not admitting of a boat's landing, owing to the rocks and kelp; on the opposite beach, in the background, is seen some small hair seal rookeries, with six or eight clapmatches (female seal) as usual huddled around each sea lion, their protectors; on the upland hillocks, the coarse tussuc grass appears, while over the more elevated ground in the interior, the albatrosses are hovering, and directly over the back of the cook, a whale-boat is seen crossing, and coming round the point with a load of skins.

For the information of these engaged in the right or black whale fishery, the author would here mention, that these monsters of the deep are to be met with, in great numbers, at the mouth of Sinfonda Bay, on the coast of Patagonia, in about latitude 42° 50' south. For several leagues along this coast to the northward, and near the shore, it is presumed that a ship might lay off and on, or even come to an anchor, near the shore, during the prevalence of the westerly winds, and take these fish as rapidly as the crew could cut them in. They were so numerous in the passage between the chops of this bay, and so gentle and void of fear withal, as in their gambols around the ship, frequently to come so near that the spray from their spouts fell upon the deck; her presence did not disturb them in the least, for they continued their sportings as if she had not been there. Many of the like species of whale also frequent the bays, inlets, sounds, and passes of the Falkland Islands, particularly in the passes between the New and Swan Islands, and between Swan Island and the Great Malone. They are to be found in still greater numbers at the south mouth of Falkland Sound, both in the west branch, by Arch Islands, and in the east branch, by Eagle Island. These islands afford good harbors, where vessels may anchor and obtain cargoes of oil and bone, as fast as

SEALER'S ENCAMPMENT, BYERS ISLAND, FALKLAND ISLANDS
From a lithograph in Fanning's *Voyages*, New York, 1833

PALMER'S LAND AS SEEN FROM THE SOUTH SHETLANDS
From a lithograph in Fanning's *Voyages*, New York, 1833

the oil can be tryed out: the whales here appear perfectly tame and quite fearless, and are not known ever to have been disturbed. A person at a station on any elevated bluff, from January to June, may observe their blows in all directions, rising high out above their element, and this is the case throughout the whole day, with seldom any cessation. They frequently feed close in shore, at the verge of the kelp, as well as at times making a circuit round in the harbors. They are also to be found in vast numbers at Nassau Bay, at the mouths of the inlets and passes near and in the vicinity of Cape Horn. A ship steering to the southward, and rounding this celebrated cape, when at a fair distance to the west, may then direct her course to the northward, up Nassau Bay, and gaining the western shore on board, will soon be brought in view of good harbors, where she can anchor in safety, to attend to her whaling business.

After procuring a full cargo of elephant oil, and upwards of 5,000 seal skins, our ship started for home. On the 22d of April, in latitude 38° 8′ north, and longitude 68° 25′ west, she was overtaken by a violent storm of wind, rain, and hail, and an almost continued flash of lightning, attended with terrific peals of thunder, the wind coming in heavy squalls, first from the northeast, then from the east and southeast, and then round to the southwest, and between each squall falling nearly to a calm, thus creating a confused cross-breaking sea; this caused the ship to labor very much, although her sails were all snugly furled, except a reefed fore-course, which it was necessary to attend all the time by the braces, on account of the sudden changes of the wind. Somewhere about 11 P. M. the lightning struck the main-mast, and followed it down to the fair leader; here two of our choicest men were hauling in the fore-braces, who in an instant were hurled into the lee waist. At this moment, the dreary darkness gave way to the bright glare of a flash, yet this, if it were possible, left all things darker as it vanished; the ship was filled, both on deck and below, with a most disagreeable sulphurous smell, somewhat like that caused by burning damaged gun powder, rendering it difficult to breathe for some minutes, though the smell remained for hours after. The two men who were wounded, were immediately taken into the cabin, and attended to by our first officer, Mr. D. Mackay, who possessed, of all others on board, the greatest knowledge in physic and surgery; they were found to be differently wounded; the eldest, who came first to his senses, had received two injuries, one on the arm near to the shoulder, which was burnt a little, and a hole perforated nearly through the thick or fleshy part of the thigh, in size like that caused by the ball of a rifle, while on the opposite side of the limb, or at the termination of the progress of the fluid, the flesh was seared quite hard, and about the size of half a dollar, resembling in look a piece of horn; the nails of the hand were also seared, and though there was scarcely any bleeding, the wound, however, was very painful. The other man was, as it were, completely roasted, from his neck down to the knees; by putting any pressure on the skin, it would snap, or rather crack, similar to that of a roasted pig;—poor fellow! it was a sad sight. His outer garments had remained untouched by the lightning, those underneath being slightly scorched, while the waistcoat was a little burned, but the guernsey robbin nearest his person, was burned black and to a tinder. Even philosophy was put to a stand, to account for so different an effect upon two persons clinging to the same rope, for when a sample of the shirt was presented to Doctor Samuel L. Mitchill for his opinion, he could not but say, that he thought it a very singular freak in the electric fluid, thus differently to act upon the two subjects.

A few drops of laudanum had been occasionally administered to relieve the pains, and his person was instantly, on being brought into the cabin, dressed in linen bandages, continually moistened with olive oil; still notwithstanding all our best exertions and attentions, it was five hours before he came to his senses; the heaving of the ship also added in no small degree to both their distress.

At 6 P. M. three days after, we hove in sight of the land, which proved to be the south side of Long Island, about midway of the same; the wind at the time blowing fresh from the W. S. W there was little prospect of our being able to fetch Sandy Hook; therefore, bore away for the east end of the island. At 8 A. M. we passed Montague light, and at noon came to anchor in the mouth of New London harbor. An officer was immediately sent to the town, who returned shortly with the health physician of the port in company, who, after examining our wounded men, advised me to send them without delay to New York, the necessary means for their recovery not being within his reach.

Got the ship under way the next morning, and proceeded as far as Whitestone; while at anchor here, waiting to receive a Sound pilot, a New Haven packet crossed the ship's stern, she having the tide in her favor, there was every prospect of her reaching the city in a couple of hours; (steam-boats were not so common then as at the present day) her captain, Benjamin Beecher, upon being made acquainted with the circumstances of the case, readily consented to take our invalids to the city, although he had upwards of forty lady passengers, besides many gentlemen; these in like manner consented to their being received on board. Too much credit cannot be given to Captain Beecher for his kind attention and exertion, to alleviate the distress and pains of our fellow-voyagers, while under his care, still suffering as they were very severely from their wounds. Many of the ladies, true to their nature and creditable to themselves, immediately surrounded the cots, which had been prepared in the cabin, and administered to the men's comfort. Aware that the law requiring a ship to be first visited by the health officer of the port, before any of her company could leave, had been now broken, a note was sent to the collector in New York, confessing the same, but pleading the urgency of the case in excuse, asking his assistance also in the procuring of medical aid in their behalf. This was most cheerfully accorded by David Gelston, Esq., the collector, who remarked, that the law never had been made for fines in such a case. In the hands of the skillful faculty of the hospital, both men shortly after were entirely cured.

The *Sea Fox* arrived at the dock in New York the

second day after, having been absent seven months and twenty-three days. The closing up of her voyage, by the sale of the cargo and vessel, produced to her owners a nett profit of eighty-eight per cent, in short of ten months from the time they advanced the principal. It is worthy of notice, that the potatoes taken on board the first day of September, 1817, kept well in the nettings between the carlings, until the last were used, about the middle of March following: during this time the ship had twice crossed the equator, and been in upwards of fifty degrees south. In the same way oranges received on board at St. Ann's Island, on our outward passage, were in good preservation when the ship again had soundings on the American coast; great care, however, was taken to pick out the sprouts and defected ones every week; by means of these articles, the health of our crew had been beyond doubt greatly preserved.

In July of 1819, the brig *Hersilia*, a fine new vessel, coppered and fitted in the best manner, sailed from Stonington, on an exploring and sealing voyage, under the command of James P. Sheffield, William A. Fanning, supercargo. In addition to his being then in possession of the corrected survey of the Spanish corvette, Atrevida's position of the Aurora Islands, also of the manuscript of Captain Dirck Gherritz's discovery of land to the south of Cape Horn, in the Dutch ship *Good News*, in the year 1599. The author had previuosly been, in the spring of the year, at South Georgia at the breaking up of the winter ice, a few days after a gale had set in from the W. S. W.; fleets of ice islands came from that quarter, and in passing to the eastward, brought up against the south-western coast of South Georgia, giving decisive evidence that extensive land did exist in that direction, for as numerous ice islands had formed at South Georgia, and drifted away to the eastward in these gales, it was certain that the ice islands first spoken of, must have had land to form at, or they could never have been in existence. The author was therefore convinced that land was to be found, somewhere between the latitudes of 60° and 65° south, and between 50° and 60° west; besides this, the correctness of the manuscript of Gherritz's discovery, was beyond a doubt.

The master and supercargo of the *Hersilia,* both possessing nautical talents, and both able lunarians, were therefore directed, in their instructions, to touch first at the Falkland Islands, there to fill up their water and refresh the crew, thence to proceed in search of the Aurora Islands, and should seals be there found, to procure their cargo, if not, to return westward to Staten Island, and after wooding and watering, to stand to the southward, keeping as nearly in the latitude of Cape Horn as the winds would admit, until they arrived in the latitude of about 63° south, then to bear up and steer east, when it was confidently expected they would meet with land; but after all, should they be still unfortunate in the search, and find no seals, then to enter the Pacific, or return to the Falklands, or islands about Cape Horn, and endeavor to procure a cargo.

On the return of the brig, they reported having on the passage out touched at the Falklands, thence proceeded in search of the Aurora Islands; these were found to be three in number, each in form of a sugarloaf, but having no landing places, even for amphibious animals, on them. A number of birds about, with some shaggs and white pigeons in the clefts, were all the living creatures discovered on them. The brig sailed around and between these islands without discovering any danger, except a reef which put off southwest, a short mile from the southernmost island. The centre island they place in latitude 52° 58′ south, longitude 47° 51′ west.

Leaving the Auroras, the brig's course was shaped westward for Staten Land; from this, after taking on board wood and water, they steered to the south; on arriving in about the latitude of 63° south, at 1 P. M. they bore away east, under a good breeze from the westward, attended with clear weather; hove to when night closed in, during which many seals came swimming about the vessel; this gave them strong hopes of being in the vicinity of land. In the morning, bore away again to the eastward, and at 10 A. M. to their great joy, a high and round mountain island was discovered covered with snow, although in the month of February, and the last summer month in this region. From its singular form they named this Mount Pisgah Island; upon approaching nearer, more land, or rather mountains, of craggy rocks, to the eastward were discovered.

After passing Mount Pisgah Island, they arrived at the group last seen, and called them Fanning's Islands; after sailing into a passage between the first two, they came to a harbor at the starboard island which was then named Ragged Island, and there anchored, calling it Hersilia cove. From elevated positions they had discovered more land to the eastward, but as the season was drawing to a close, and they were anxious to hasten home and report the discovery of such vast numbers of seals to their friends in time for the next season, they had no leisure to visit or make a survey of it. After procuring several thousand skins of the choicest and richest furs, as the weather or climate would not admit of their drying them at this place, they were therefore not prepared with a sufficiency of salt for a full cargo, but calculated to dry a part of the skins where they should take them; thus with as many as they had salt to save, they left Ragged Island, leaving, according to their estimation, 50,000 fur seals. The *Hersilia* returned safely to Stonington, and realised to her owners a very handsome profit upon the sale of the vessel and her cargo.

These islands had been seen, it appears, by a Captain Smith, in the English brig *William,* some fifteen months prior to the arrival of the *Hersilia* at them, while bound from Buenos Ayres to Valparaiso, and by him named the South Shetlands. But as Captain Gherritz, after his discovery of this land in 1599, proceeded into the Pacific, and was there captured by the Spaniards and sent into Valparaiso, it is therefore likely that Captain Smith, who was then engaged in the freighting trade to Valparaiso, got a hint of the Dutch captain's discovery from some Spanish source, and then went to look after it. Succeeding in his search, he then claimed it as a new discovery, and gave it an English name, by which it is established and recorded on the charts in use, the public generally not being aware that Captain Gherritz, was the first discoverer; it would have been more liberal and just in him to have named them D. Gherritz's Islands. We Yankees might with more propriety, inasmuch as they had become obsolete, have considered our rediscovery of the Crozetts, after the long and tedious search had for them, as a discovery, and named them South Martha's Vineyard, or something else. The *Hersilia*, the first American vessel it is believed that visited, did not presume to call the group or chain by any other name than that given by their first discoverer, viz. Gherritz's New Iceland. Captain Gherritz states his having sailed many miles along its coast, and that it consisted of high, craggy, sterile mountains, like to the coast of Norway, and was covered with ice and snow. The author does not intend, however, to enter into any disputation touching the names.

The South Shetland Islands, now so called, are a chain of rough, rocky, and mountainous islands, whose valleys or chasms are partially filled with everlasting ice, and during the greatest part of the year they are covered with snow. The chain consists of upwards of fifty islands and islets, stretching in a southwest and northeast direction, and are situated in latitudes 61 and 63 and a half degrees south, and between 54 and 63 degrees west. The navigation among the group is dangerous on account of many sunken reefs. The weather is similar to that of South Georgia; there is very little earth or vegetation, except the winter moss, to be seen, nor is there a bush or shrub to be found. Deception Island, the most southerly, is a curious phenomenon of nature, and beyond doubt of volcanic origin: in form it is a mountain ridge, making the interior round the bay in appearance an immense bowl, while on the east side as it were, is a piece broken out; this forms an opening or passage by which vessels enter the bay. At the northeast inner bay side, is the harbor called Yankee Harbor, near to which, along the shore, is a stream of hot or boiling water; this keeps the water of the bay, for a little distance round, quite warm, and is much resorted to by the disabled and wounded penguins, who appear fond of, and anxious to remain in it. By scraping down a few inches into the sand of the beach, a few yards distant from the boiling fount, the heat is so great as to render it impossible to hold the hand in any length of time; notwithstanding, very near by, in the cavity of the mountain, is an iceberg of solid flint ice, several hundred feet in heighth.

The next season after the *Hersilia's* return from the South Shetlands, a fleet of vessels, consisting of the brig *Frederick,* Captain Benjamin Pendleton, the senior commander, the brig *Hersilia,* Captain James P. Sheffield, schooners *Express,* Captain E. Williams, *Free Gift,* Captain F. Dunbar, and sloop *Hero,* Captain N. B. Palmer, was fitted out at Stonington, Connecticut, on a voyage to the South Shetlands. From Captain Pendleton's report, as rendered on their return, it appeared that while the fleet lay at anchor in Yankee Harbor, Deception Island, during the season of 1820 and 21, being on the lookout from an elevated station, on the mountain of the island during a very clear day he had discovered mountains (one a volcano in operation) in the south; this was what is now known by the name of Palmer's Land; from the statement it will be perceived how this name came deservedly to be given it, and by which it is now current in the modern charts. To examine this newly discovered land, Captain N. B. Palmer, in the sloop *Hero*, a vessel but little rising forty tons, was despatched; he found it to be an extensive mountainous country, more sterile and dismal if possible, and more heavily loaded with ice and snow, than the South Shetlands; there were sea leopards on its shore, but no fur seals; the main part of its coast was ice bound, although it was in the midsummer of this hemisphere, and a landing consequently difficult.

On the *Hero's* return passage to Yankee Harbor she got becalmed in a thick fog between the South Shetlands and the newly discovered continent, but nearest the former. When this began to clear away, Captain Palmer was surprised to find his little barque between a frigate and sloop of war, and instantly run up the United States' flag; the frigate and sloop of war then set the Russian colors. Soon after this a boat was seen pulling from the commodore's ship for the *Hero,* and when alongside, the lieutenant presented an invitation from his commodore for Captain Pendleton to go on board; this, of course, was accepted. These ships he then found were the two discovery ships sent out by the Emperor Alexander of Russia, on a voyage round the world. To the commodore's interrogatory if he had any knowledge of those islands then in sight, and what they were, Captain Pendleton replied, he was well acquainted with them, and that they were the South Shetlands, at the same time making a tender of his services to pilot the ships into a good harbor at Deception Island, the nearest by, where water and refreshments such as the island afforded could be obtained; he also informing the Russian officer that his vessel belonged to a fleet of five sail, out of Stonington, under command of Captain B. Pendleton, and then at anchor in Yankee Harbor, who, would most cheerfully render any assistance in his power. The commodore thanked him kindly, "but previous to our being enveloped in the fog," said he, "we had sight of those islands, and concluded we had made a discovery, but behold, when the fog lifts, to my great surprise, here is an American vessel apparently in as fine order as if it were but yesterday she had left the United States; not only this, but her master is ready to pilot my vessels into port; we must surrender the palm to you Americans," continued he, very flatteringly. His astonishment was yet more increased, when Captain Palmer informed him of the existence of an immense extent of land to the south, whose mountains might be seen from the mast-head when the fog should clear away entirely. Captain Palmer, while on board the frigate, was entertained in the most friendly manner, and the commodore was so forcibly struck with the circumstances of the case, that he named the coast then to the south, Palmer's Land;* by this name it is recorded on the recent Russian and English charts

* This continent, it is asserted in Morrell's Voyage, page 69, was named "New South Greenland," by a Captain Johnson. It is but just to state here, that this most meritorious mariner (Captain Johnson) was a pupil to, and made his first voyage to the South Seas with, the author, with whom also he remained, rising to different stations, and finally became one of his best officers; the first information he obtained of the discovery of this land by Captains Pendleton and Palmer was from the author of this work.

and maps which have been published since the return of these ships. The situation of the different vessels may be seen by the plate; they were at the time of the lifting of the fog and its going off to the eastward, to the south, and in sight of the Shetland Islands, but nearest to Deception Island. In their immediate neighborhood were many ice islands, some of greater and some of less dimensions, while far off to the south, the icy tops of some two or three of the mountains on Palmer's Land could be faintly seen; the wind at the time was moderate, and both the ships and the little sloop were moving along under full sail.

The following season, in 1821 and 22, Captain Pendleton was again at Yankee Harbor, with the Stonington fleet; he then once more despatched Captain Palmer in the sloop *James Monroe,* an excellent vessel of upwards of 80 tons, well calculated for such duties, and by her great strength well able to venture in the midst of and wrestle with the ice. Captain Palmer reported on his return, that after proceeding to the southward, he met ice fast and firmly attached to the shore of Palmer's Land; he then traced the coast to the eastward, keeping as near the shore as the ice would suffer; at times he was able to come along shore, at other points he could not approach within from one to several miles, owing to the firm ices, although it was in December and January, the middle summer months in this hemisphere. In this way he coasted along this continent upwards of fifteen degrees, viz. from 64 and odd, down below the 49th of west longitude. The coast, as he proceeded to the eastward, became more clear of ice, so that he was able to trace the shore better; in 61° 41' south latitude, a strait was discovered which he named Washington Strait, this he entered, and about a league within, came to a fine bay which he named Monroe Bay, at the head of this was a good harbor; here they anchored, calling it Palmer's Harbor. The captain landed on the beach among a number of those beautiful amphibious animals, the spotted glossy looking sea leopard, and that rich golden colored noble bird, the king penguin; making their way through these, the captain and party traversed the coast and country for some distance around, without discovering the least appearance of vegetation excepting the winter moss. The sea leopards were the only animals found; there were, however, vast numbers of birds, several different species of the penguin, Port Egmont hens, white pigeons, a variety of gulls, and many kinds of oceanic birds; the valleys and gulleys were mainly filled with those never dissolved icebergs, their square and perpendicular fronts several hundred feet in height, glistening most splendidly in a variety of colors as the sun shone upon them. The mountains on the coast, as well as those to all appearance in the interior, were generally covered with snow, except when their black peaks were seen here and there peeping out.

The schooner *Pacific,* Captain James Brown, sailed on a sealing voyage to the South Seas, from Portsmouth, October 1st, 1829. November 14th she reached the Cape de Verds, and there took in salt and fresh provisions, and sailed for South Georgia, in the South Atlantic, which she made on the 29th December, 1829. Left Georgia March 5th, with 256 skins, and 1,800 gallons sea elephant oil on board. December 8th, 1830, latitude 56° 18′ south, longitude 28° 35′ west, they discovered an island neither laid down on any chart, nor described by Cook or Bowditch. In clear weather, the land may be seen thirty miles off; the island is two miles in circumference, resembling at a distance a high round lump; Captain Brown named this Potter's Island.

Four days after, viz. on the 12th, a second island was discovered, having a mountain 800 feet high in the centre, from which, in several places, smoke was constantly issuing; it was covered with ice and snow; on the lower or level part of the island was a deep stratum of lava disgorged from the volcano; this was of a light brown color, and somewhat resembling brick when burnt to excess, extremely porous and fragile, and so light as to float on the water; on the ocean in the vicinity of this island, large masses of this were seen floating about. The seamen went on shore, and travelled over several portions of the island, examining the places whence issued the smoke; at these a slight degree of heat was perceptible; upon digging down in the earth several feet below the surface, the ground was found to be extremely dry. On this isolated spot are two stony beaches, and convenient landings. Five different species of the penguins were found here, as also nelleys, spotted eaglets, sea hens, gulls, &c. in great abundance. This they named Prince's Island; it is five miles long from N. W. to S. E. and lies in latitude 55° 55′ south, longitude 27° 53′ west.

22d. Another island, and lying in latitude 56° 25′

south, longitude 27° 43' west, was discovered, being six miles in length from N. W. to S. E.; there was also on this a burning mountain, with smoke issuing in different places; it has no landing, and may be seen in clear weather fifty miles. Captain Brown named this Willey's Island. The fourth and last that Captain Brown discovered, is situated in latitude 57° 49' south, longitude 27° 38' west. It received the name of Christmas Island, having been first noticed on the 25th of December, 1830, and lies midway between Candlemas and Montague Island, but farther westerly than either, as laid down on the chart by Mr. Prince, Captain Brown's mate, an experienced seaman, who traversed these waters in an English vessel twelve years ago.

The largest icebergs seen by the Pacific, were in latitude 58° 18' south; some of these were three to four miles in length, two in breadth, and from two to three hundred feet high, and flat on the top. The coldest weather was in June, July, and August; the hottest in December and January. On the newly discovered islands there was neither wood, timber, nor vegetables of any kind. On Bird's Island, the crew of the *Pacific* killed a sea tiger, measuring eighteen feet; the skull and hide of one of these animals was brought home by them; the animal was seven feet in length, and girted three and a half when killed. The head is shaped like that of the common seal, except that is more elongated, the sockets of the eyes too, being deeper and broader; it measures fifteen and a quarter inches from the extremity of the nose to the great hole of the occipital bone; the lower jaw, from the chin to the point of articulation with the upper jaw bone, is eleven and a half inches. A straight line drawn from one articulating process to the other, measures six inches. The number of teeth is thirty-two, four of which are tusks; the largest of these is an inch and a quarter in length, and one in circumference at the base; in each jaw were ten grinders; these immediately after emerging from their sockets, are divided into three distinct conical portions, the central ones being more than half an inch long, and the other two the fourth of an inch, all terminating with sharp points. The skin is covered with a thick, fine, and short hair, on the back of a gray color, spotted with black, and white on the abdomen; the flippers are short and strong; the animal moves with surprising velocity in the water, and in that element all its motions are indicative of great strength; their chief food consists of penguins. To catch these beautiful birds, when they are discovered at a distance, the tiger gets upon the windward side, and lies upon his back; in this position he floats upon the billows, with his head a little elevated, but all the while keeping his dark vigilant eye steadily fixed upon the ill fated object of his pursuit; as soon as he is sufficiently near to secure his prey, he turns suddenly upon his belly, cleaves the billows with astonishing swiftness, and the next moment is seen plunging in the water with a penguin, weighing at times from forty to sixty pounds, in his capacious jaws.

The tiger possesses undaunted courage and shrewdness; they frequently chased the crew of the *Pacific* while cruising in their boats. On one occasion, when two of the men were at a considerable distance, both from the shore and schooner, they were discovered by one of these animals, some twenty feet in length, and six in circumference, which instantly pursued the boat with all speed, and when within ten or twelve feet, leaped for it, exposing to view at the same time in the greatest rage, his sharp teeth. Failure in this attempt, he next essayed to upset the boat; one of the party then lodged a ball in his body; this only served to increase the animal's rage, and in another attempt to spring into the boat, he would have succeeded, but for a severe blow he received from a lance. Even after this, his courage and perseverance were unabated, and it seemed as if he had resolved that neither the power nor the weapons of man should prevail against him. When, however, the second and third balls were lodged in him, his efforts ceased, and he was overcome.

On another occasion, some of the crew were in the boat three miles from the schooner, when a large tiger was observed following in their wake; he betrayed no disposition to annoy them, but kept at a distance from the boat all the time; the seamen, unacquainted with his cunning, were induced to pursue him, but soon found their ignorance of the animal's character had betrayed them into very imminent danger, which they were now likely to pay very dearly for; the tiger waited their approach, and then commenced the battle, when the seamen instantly retreated for their vessel, and with the utmost difficulty succeeded in keeping him from upsetting them.

Some of the sailors tasted the milk of the sea tiger which they had killed, and found it excellent. By many persons it is supposed that the sea tiger and walrus are the same; but they differ in several particulars, such as the number, size, shape, and relative position of the teeth, as well as in the form of the head, which of the walrus bears a strong resemblance to that of the human species.

CHAPTER XXI

MODERN ASSERTED DISCOVERIES

Pike's Island, latitude 26° 19' south, longitude 105° 16' west, discovered in 1809.

Ducie's Island, latitude 24° 26' south, longitude 124° 37' west.

Mitchill's Group, latitude 9° 18' south, longitude 179° 45' east, discovered by Captain Barrett, in the ship *Independence,* of Nantucket. This group is inhabited.

Rocky Island, latitude 10° 45' south, longitude 179° 28' east, variation 11° east, discovered by Captain Barrett, of Nantucket.

Swain's Island, latitude 59° 30' south, longitude 100° west, by calculation, discovered by Captain Swain, of Nantucket, in 1800. Resorted to by many seals.

Tuck's Island, latitude 17° north, longitude 155° east. Very low, and inhabited.

Worth's Islands, latitude 8° 45' north, longitude 151° 30' east, five in number.

Tuck's reef and sail rocks, nine in number, latitude 6° 20' south, longitude 159° 30' east.

Rambler's reef, latitude 21° 45' north, longitude 175° 12' east.

Rambler's reef, latitude 23° 29' north, longitude 178° 13' east.

Rambler's reef, latitude 23° 30' north, longitude 178° 31' east. These, from Tuck's Island, were all discovered by Captain William Worth, second in the *Rambler*, of Nantucket, in 1829.

Jefferson's Island, latitude 18° 27′ north, longitude 115° 30′ west, discovered by a ship out of Salem, April 8th, 1826.

Gardner's Island, latitude 4° 30′ south, longitude 174° 22′ west.

Coffin's Island, latitude 31° 13′ south, longitude 178° 54′ west.

Great Ganges Island, latitude 10° 25′ south, longitude 160° 45′ west. Inhabited.

Little Ganges Island, latitude 10° south, longitude 161° west. Inhabited, and affording cocoa-nuts, &c. These four last mentioned were discovered by Captain J. Coffin, in the ship *Ganges*, out of Nantucket. The natives were friendly, and readily brought off to the ship, cocoa-nuts, &c.

Unknown Island, latitude 5° south, longitude 155° 10′ west, about ten miles long, and two wide; rocky shore.

Reaper Island, latitude 9° 55′ south, longitude 152° 40′ west. Low, woody, and uninhabited. Discovered by Captain Coffin, in 1828.

Group Islands, latitude 31° 25′ south, longitude between 129° 27′ and 130° 15′ west, discovered by Captain J. Mitchel, in 1823.

Lancaster reef, latitude 27° 2′ south, longitude 146° 27′ west, tending six miles N. E. and S. W. discovered by Captain Weeks, of New Bedford, 1830.

Oeno Island, latitude 23° 57′ south, longitude 131° 5′ west, about eighty miles N. W. by N. of Pitcairn's Island. A dangerous reef puts out from the south point. Discovered by Captain G. B. Worth, in the ship *Oeno,* of Nantucket.

Unknown reef, latitude 27° 46' north, longitude 174° 56' west, rocks above water, with sand bars, where the ship *Pearl,* Captain Clark, and *Hermes,* Captain Phillips, were wrecked, April 26th, 1822. The crews were saved, and taken off, after remaining two months on the reef.

Smut-face Island, latitude 6° 16' south, longitude 177° 19' east.

Parker's Island, latitude 1° 19' south, longitude 174° 30' east.

Brown's Island, latitude 18° 11' south, longitude 175° 48' east. These three last mentioned islands were discovered by Captain Plasket, in the ship *Independence,* of Nantucket, in 1828.

Chase's Island, latitude 2° 28' south, longitude 176° east.

Lincoln's Island, latitude 1° 50' south, longitude 175° east.

Brind's Island, latitude 0° 20' north, longitude 174° east.

Dundas Island, latitude 0° 10' north, longitude 174° 12' east. These four last mentioned islands were discovered by Captain Chase, in the ship *Japan,* of Nantucket, in 1827 and 1828.

Nixon's rock, latitude 40° south, longitude 57° 36' west, six feet above water, tending N. E. a cable's length. Discovered by Captain Dixon, in the *Ariel.*

New Discovery Island, latitude 15° 31' south, longitude 176° 11' east. Inhabited, and discovered by Captain Hunter, in the *Carmelite.*

Valetta Island, latitude 21° 2' south, longitude 133° 13' east, discovered by Captain Philips, in the *Valetta*, July 10th, 1825.

Whale rock, latitude 51° 51' south, longitude 64° 32' west, just above water, with much kelp attached to it.

Gardner's Island rock, latitude 25° 3' north, longitude 167° 40' west, about one mile in circumference, and one hundred and fifty feet in height.

Allen's reef, latitude 25° 28' north, longitude 170° 20' west. These were both discovered by Captain J. Allen, in the ship *Maro*, of Nantucket, in 1821.

Starbuck's Group, latitude on the equator, longitude 173° 30' east.

Loper's Island, latitude 6° 7' south, longitude 177° 40' east.

Dangerous reef, latitude 5° 30' south, longitude 175° west.

Tracy's Island, latitude 7° 30' south, longitude 178° 45' east.

New Nantucket, latitude 0° 11' north, longitude 176° 20' west.

Granger's Island, latitude 18° 58' north, longitude 146° 14' east. These six last mentioned were discovered by Nantucket whale ships, from 1820 to 1826.

Fisher's Island and Group, latitude 26° 30' north, longitude 141° 1' east, discovered by the British ship *Transit*, Captain J. J. Coffin, September 12th, 1824.

Captain Coffin reports, that he found this group to consist of six islands, besides a number of rocks and reefs. The largest he called Fisher's Island; the second in size, Kidd's Island, after his owners in Bristol, England; the third, being the most southern of the group, he called South Island; the fourth, from the abundance of pigeons found thereon, he named Pigeon Island. About four miles E. N. E. of South Island, lie two high, round islands, to which the captain gave no names. Fisher's Island is about four leagues in length, tending S. S. E. and N. N. W. Kidd's Island, the most western of the group, lies S. E. from the N. W. part of Fisher's Island. Between the two last mentioned islands, is a beautiful clear bay, two miles wide, and five miles up to the head. The Transit sailed up the bay about four miles, where near to Fisher's Island, a fine small bay was found; to this Captain Coffin very properly gave the name of Coffin's Harbor. It is sheltered from all winds, except the W. S. W., and has no current or swell; here, too, he anchored his ship in fifteen fathoms of water. In three days the captain took on board of his ship fifty tons of water, of the purest quality, and a sufficient supply of wood; both of these very essential articles he found very abundant, and more easily procured than at any other place he was ever at. Turtle and pigeons were also so plentiful that any quantity could be obtained, a limitation for the number to be taken daily being laid by Captain Coffin, to prevent a useless waste. The bay is well stored with a variety of excellent fish, and plenty of lobsters. Among the productions of the island, is the cabbage tree, which can easily be obtained in any quantity. No quadruped, reptile, or insect of any kind, not even an ant, was discovered by Captain Coffin.

The islands are covered with large and beautiful forest trees; on any of these there was no mark, made by a knife or otherwise, traceable, by which it could be made to appear that man had ever been on any of these islands. For ships employed in the whale fishery, or bound from Canton to Port Jackson, or the north-west coast of America, they furnish a desirable place for refreshment. These islands are about south of Sandown Point, on the coast of Japan; the distance may be sailed in four days. The latitude and longitude as above given, is for the harbor where Captain Coffin anchored, and took in his wood, water, &c.

Covell's Group, consisting of fourteen islands, lies in latitude 4° 30' north, longitude 168° 40' east. Discovered by Captain H. Covell, in the bark *Alliance*, May 7th, 1831. The group is inhabited.

CHAPTER XXII

SANDAL WOOD, BEACH LA MER, &C.

SANDAL Wood. This valuable wood is thought to be a native plant within the torrid zone, or growing solely under the 25th degree of latitude from the equator. It grows only in a mountainous country, and most generally on an elevation above the second region in the mountain range. It was formerly procured from the peninsula of India, but for the last twenty years, many cargoes, principally by American enterprise, have been taken on board at the Fee Jee and other mountainous islands in the Pacific. At these places the wood is considered to be the sole property of the king or chief, and whenever a cargo is wanted at an inhabited island, they are the persons to make the bargain with; it is also found on uninhabited islands in this ocean. To procure it, the laborers taking with them the axe, cross-cut saw, drawing-knife, spade, and grub-hoe, the necessary tools, proceed up the mountain to where it grows; to prevent waste the tree is sawed down near to the ground, and then cut into lengths of about four feet; all the smaller limbs of a quarter of an inch in diameter at the smallest end are also taken. After this, the bark and sap must be nicely shaved off, before the wood is in a merchantable state for market. Whenever it is necessary to proceed in the most saving manner on account of there not being plenty of trees, they can be felled by attaching a line at the top, and clearing away the sod and earth from off the roots (these and the stumps are worth within twenty per cent. of the body and limbs), some four feet or so, then after cutting the roots at this distance, the tree may be pulled down, and then sawed and shaved for the market. There are two kinds of the wood, a genuine and spurious, both similar in size and form to the apple tree; they differ so little in the bark, leaf, and fragrance, at the time the tree is felled, that it requires the judgment of an experienced person to select the genuine. The spurious kind in a very short time loses its fragrance, and then is of no other value than for fire wood; instances have occurred of its having by mistake been taken to Canton, only from the want of capacity to judge in the person who made the selection; this had obliged the importer, in addition to his disappointment, to throw the wood into the river during the night to avoid the necessity of paying duties on it; the genuine on the contrary retains its oil and fragrance for many years. Trees in full vigor and growth are the best, and produce the first chop wood in the Canton market; if the tree is old, it is not only apt to be defective, but much worm-eaten, thereby greatly diminishing its value, and if too young they are not so fragrant. The China merchants make a difference in estimating their value of from 20 to 70 per cent between the very first chop (as they call it) and the third, or fourth, or last chops. This wood is highly impregnated with an essential oil of great sweetness of smell, and in China the chief market for it, it is greatly esteemed as a sacred wood, and as such offered by the Chinese in sacrifice to their god Josh, and for this reason called Josh wood. Notwithstanding this, the wealthier persons have pieces of furniture made wholly or in part of it; the perfume with which it fills those rooms wherein it is placed, being considered a great luxury by all classes who desire to have every article about them tinged therewith; out of it the Chinese manufacture fans and other articles, but the highest value they attach to it is as their sacrifice to Josh. These offerings are kept constantly burning on the altars before the image at their houses of worship, the priests carefully attending to, and renewing the fire as often as is necessary by placing several small pieces on their ends on the top of the altar, similar to the stacking of muskets. As they assert, and apparently believe, that Josh would destroy them all were they to be without a supply of this wood, they therefore make great exertions to procure it at any price.

With the oil that the priests extract from this wood they prepare little square sheets of gilt edged paper called Josh paper; this is disposed of in packs to the heads of families, so that when sickness or any other trouble visits the family or any of its members, the head of the same takes one of these sheets, and rolling it up as a sheroot or segar, sets one end on fire, and as it burns, waves it about with the hand to obtain favor in their distress; this is called chin-chin-ing to Josh (sacrificing to their God).

In searching for the sandal wood on those islands situated in the most western part of the Pacific Ocean, wild nutmeg-trees have been found in groves, or rather intermixed with the forest trees. The nut is of greater length than the domestic nut, and of but little value, being in truth not worth gathering, for when plucked it is weak in richness and spiciness of flavor, and therefore soon loses even the little of this it ever had. The tree appears to grow longer and ranker than the common nutmeg-tree.

Eatable bird's nests. This article is the nest of a small bird, of a half circle in form, and very similar to the barn swallow's nest, though not so large or bulky by at least the one third; the nest is of a gummy thread about the size of sewing twine, the better kind being of a white or bright and clear amber color, the yellow and black are of much less value. In all probability it has not yet been discovered where or from what the bird procures this substance, many supposing it to be from the gum of some tree; the author is of opinion, however, that it is a marine production, as when cooked, the taste, an epicure will discover, to be somewhat similar to the marine coral moss. The nests are attached to the clefts in the rocks on the seacoast, and to gather them, the boat should be one of a buoyant kind, in a rough sea or surge; in this the collecting hands should be in pairs, one with a small hook something similar in shape to a bill-hook, only more hooked at its point, in length six or eight inches, well secured to a pole ten or eighteen feet long, with which the nests can be hooked off, while his companion with a scoop-net a little bagging, about the size of, or larger than, a barrel head, catches them as they fall; a quantity being thus collected the feathers and dirt are carefully removed, the nests then after being thoroughly dried in the sun are packed in boxes for the market. Great care should be first taken to line the boxes with dry mats or thick paper; the nests should by all means be kept from being wet or sprinkled with the spray of the sea when collected, as in this case it will be difficult to preserve them for market, for if again dried, and to all appearance perfectly so, before packed, they will, nevertheless, in a short time after they are stowed in the vessel's hold, begin to give, become moist, soft, mouldy, and spoil, or at least be considered as damaged, and not prime. The Chinese highly prize this article for their soups and stews, though from its rarity and high price, only the nobles and wealthy can often afford a dish of eatable bird's nest for their table.

Beach la mer. This is a species of marine worm, or if it may be so expressed, animal fish, of a glutinous nature. It is found on the coral banks and reefs, generally within the torrid zone, in the Pacific and Indian Oceans, and adjacent seas. In form and shape it is much like a large field worm, being from six to eighteen inches in length when taken, with a narrow flat on the under side which it rests or moves on upon the reef. It is a very delicate article, for if exposed to the sun's rays, the beach la mer will melt, and in a short time dissolve away to but a mere stringy substance, and should therefore be kept very carefully covered from the sun until it has gone through a dressing process, and is in readiness for drying. The manner of collecting and curing it is to have a small punt or scow (the less water the craft draws the better) of about fourteen feet in length, turned up, scow fashion, at each end, three and a half feet wide, and two deep, covered with a canvass awning, the entire length, with drapery eighteen inches deep at the sides, so that the heat of the sun may effectually be kept from it. In this punt, two men, one at each end on opposite sides, paddle off at half ebb tide from the vessel at anchor, for the fishing bed of beach la mer on the coral reef. Arrived there, the men still at each end get out, and while wading along pick up the beach la mer in from one to three or more feet depth of water, placing the same very carefully in the punt, in this way proceeding until she is full, or until half flood, when the water becomes too deep, or the surf coming in, obliges them to break off from the reef; the men then proceed to the landing with whatever they may have obtained, where protected from the sun's rays, they prepare to dress the article by making an incision with a knife the whole length on the flat side, then taking out the entrail, for there is but one, extending its length, containing sand, with particles which in appearance resemble the coral moss; after this, being gently washed, it is immersed in a boiling caldron of strong pickle, with a small matter of saltpetre; a mite of allum added to the pickle will help to give the beach la mer a more clear amber color, which is one of the main points that recommend it at the China market; when sufficiently scalded, which experience must decide, it is taken to the flakes to be dried; these flakes are erected by means of crotches, having poles and cross-pieces running horizontally, the whole being covered with bramble twigs; this at an elevation of two feet or thereabout from the ground, allows the breeze to pass freely about them. While drying on these flakes, the smoke from burning chips, saw dust, &c. will not only assist in drying, but add to their value. It may also be moderately smoked in a house after being dried and cured in the sun, or by the fire, which answers nearly as well, and after being thus carefully dried and cured, it is in a condition to be packed in boxes or casks for market. Care should be exercised that after being cured it does not get wet, either by the salt or fresh water, as their arrival then at market in a prime and sound state may be prevented. There are three kinds or species of the beach la mer, these are designated by the Chinese as the black, yellow, and white; it is the first mentioned which is most highly esteemed, and bears the highest price among them for their soups and stews. It is very necessary that a person who undertakes to collect it should have had experience in the business, and be well able to distinguish the different kinds; if not well cured the value is lessened, and as to the white, although the Chinese will purchase it, yet it is at a price which will not pay the expense of procuring it.

The marine vegetable rock herb, or coral moss. This is a vegetable collected from the coral reefs, or rocks, and esteemed by the Chinese so highly as to be considered a royal dish, being, nevertheless, exceedingly unpalatable to European or American tastes. In both smell and taste, when served up on the table, it is like a vessel's bilge water, immediately creating nausea, and an aversion to it; neither does it possess the richness of the eatable bird's nest, or beach la mer. The Chinese, in their mode of cooking it, first parboil a piece of corned pork, which is then placed in the centre of a pan or dish, and covered over and around with this moss, several inches thick; the whole is then taken to the oven and well baked, after which, being then very brittle, it is served up on the table: when the pork is cut up finely, and mixed with the vegetable, it is then considered as a dish ready for the chop sticks of the wealthy and great among the Chinese. The moss is collected from the reef or rocks at near low water, then picked over and scalded in a cauldron of boiling liquor of brine, washed clean of the sand, &c., and when well dried, packed in mats or baskets for market; if rightly cured, and of first chop, the moss will be of a light brown color, and clear as amber: the author is of opinion, that the substance of which the eatable bird's nests is composed, is from this moss.

Mother-of-pearl. In procuring this article we have managed, by first proceeding to the pearl banks and there anchoring, after receiving on board from a friendly island in the Pacific, a number of native divers; in this place, the oysters lie at the bottom in from three to more fathoms of water; provided with a wicker basket, in which is placed ten or twenty pounds of stone ballast, as the depth of water may require, and to which the end of a coil of whale warp is made fast; the native dives, the weight in the basket assisting to carry him quickly to the bottom, where he picks up the pearl oyster, taking out the stones as the basket fills up; as soon as a sufficient weight of the oysters is in, or he wishes to come to the surface, he gives the signal by the line, and is quickly hauled up by those who are in attendance on the vessel's deck for this purpose; the diver is then allowed some little time to recover himself for another attempt. The length of time they are able, by being accustomed to it, to remain under water, is truly astonishing, varying from twenty to thirty minutes; they also take great pride in striving to outdo each other in these duties, as well as in remaining under the water; still it certainly must injure their constitutions, for on attaining the deck again, their eyes appear to have started from the sockets, while the blood ofttimes oozes from the mouth, nose and ears.

After being scraped, scrubbed, and washed clean, the oysters are spread upon the deck or ground, where exposed to a hot sun, they in a short time die; the oyster is then taken out and minutely examined for the pearls, and the mother, which is the shell, is dried, and then packed in casks or baskets, or stowed in bulk in the vessel's hold. While prosecuting their search for the oysters at the bottom, the divers will collect a reasonable proportion of turtle shell, as they occasionally catch the turtle under the rocks. After the vessel has finished this fishery, she returns with the divers to their native isle, and settles with, and pays the chief, or themselves, as has been previously agreed upon, for their time and services. The Chinese merchants rate the pearls in five chops; the first chop being the largest, of the purest water, and perfectly free from blemish; this is of the greatest value, the other four being valued proportionably down to the fifth.

INDEX

Albatross, 55.
Allen, Capt. J., 319.
Allen's Reef, 319.
Alliance *(bark)*, 321.
Anjier, 253.
Ariel *(ship)*, 318.
Aspasia *(ship)*, 206, 216, 224, 229.
Aurora Islands, 302.

Barrett, Captain, 316.
Beach la mer, 326.
Beauchene Island, 260.
Beecher, Capt. Benjamin, 300.
Betsey *(brig)*, 4, 13, 41, 42, 195.
Biddle, Captain, 294.
Bird's nests, 325.
Blanche *(frigate)*, 36.
Boca Tigris, China, 186.
Borders Island, 232, 234.
Borgues, Ignacio, 278.
Boston, J., 234, 236.
Bowne & Eddy, 14.
Brazil, 209.
Brind's Island, 318.
Brintnall, Caleb, 43, 157.
Brothers *(schooner)*, 168.
Brown, Capt. James, 310.
Brown's Island, 318.
Brumley, Capt. R., 240.
Buttersworth *(ship)*, 96.
Byers, James, 270, 295.

Canton, China, 4, 40, 189-194, 225-227, 240, 323.
Cape de Verde Islands, 5, 46.
Cape Horn, 65, 68, 223.
Cape of Good Hope, 202, 246.
Carmelite *(ship)*, 318.
Catharine *(ship)*, 242.
Centurion *(British 64)*, 228.
Chase, Captain, 318.
Chase's Island, 318.
Chinese Josh, 191.
Christmas Island, 312.
Clapmatches, 259.
Clark, Captain, 318.
Cleopatra *(English frigate)*, 207.
Cleopatre *(French frigate)*, 22.
Cod fishing, 219.
Coffin, Capt. J., 317, 319.
Coffin's Island, 317.
Coquimbo, Chili, 268.
Coral moss, 328.
Covell, Capt. H., 321.
Covell's Group, 321.
Cracatoa, 199.
Crook, Rev. William Pascoe, 89.
Crossing the Line, 6.
Curacao, W. I., 33, 34.
Cutler, B. S., 291.
Crozett Islands, 230, 242, 246, 247, 250.

332 INDEX

Deception Island, 305, 308.
Desertion of sailors, 15.
Dixon, Captain, 318.
Dolly *(schooner)*, 33, 36.
Dolphin poisoning, 50.
Doyle, —, 237.
Ducie's Island, 316.
Duff *(missionary ship)*, 89, 91.
Duke of Portland *(ship)*, 237.
Dunbar, Capt. F., 306.
Dundas Island, 318.

Express *(schooner)*, 306.

Falkland Islands, 7, 54, 60. 260, 265, 289, 295.
Falmouth, Eng., 18.
Fanning, Edmund, first voyage, 1; voyage to West Indies, 2; voyage to Falkland Islands after seals, 4; harpoons a shark, 7; voyage to England, 14; captures deserters, 15; befriended by Lord Mayor of London, 18; crew impressed, 22; obtains release of crew, 24; gale knocks ship on beam ends, 30; commands schr. *Dolly* in voyage to West Indies, 33; taken a prize by a British frigate, 37; commands brig *Betsey* on a voyage to South Seas, 41; cabin boy falls overboard, 51; reaches Falkland Islands, 54; passage round Cape Horn, 65; reaches Massafuero, 73; refits his vessel while at sea, 81; reaches the Marquesas, 84; rescues a missionary, 89; escapes from native attack, 94; visits Nuggoheeva Island, 108; goes ashore and visits the native king, 124; narrow escape from shipwreck, 156; discovers Fanning's Island, 157; discovers Washington Island, 161; miraculous escape from shipwreck, 163; arrives at Tinian Island and rescues a shipwrecked crew, 170; comes to anchor in Macao road, 183; visits Canton, 189; describes Chinese temple, 192; sails from China, 195; encounter with Malay pirates, 195-200; ship catches fire, 201; arrives at New York, 203; sails in the *Aspasia* on a sealing voyage, 206; followed by British frigate *Cleopatra*, 207; visits Pernambuco, 209; searches for Saxenburgh Island, 211; midshipman Sheffield lost overboard, 212; an ice formation, 214; reaches South Georgia, 215; builds shallops, 217; experiences with ice islands, 220; sails for Valparaiso, 222; reaches Canton and visits the city, 225; exchanges shot with the *Centurion* in the Borneo Sea, 228; reaches New York, 229; sends out the brig *Union*, 230; sends a ship to rescue Capt. Pendleton, 240; sails in the *Volunteer* for the South Seas, 265; puts into Coquimbo, Chili, and is placed in prison, 268; released and reaches New York, 293; sails in the ship *Sea Fox* for the South Seas, 295; ship struck by lightning, 298.

INDEX 333

Fanning, Gen. Edmund, 18, 37.
Fanning, Capt. Henry, 242.
Fanning, William A., 301.
Fanning's Island, 160, 247, 303.
Faulkner, Captain, 36, 39.
Fiji Islands, 234, 239, 322.
Fire, Ship on, 201.
Fisher's Island, 319.
Fogo, Island of, 5.
Fox — *(American consul)*, 22.
Frederick *(brig)*, 306.
Free Gift *(schooner)*, 306.
Freycenet, Commodore, 267.
Fur seals, 260-264, 295, 304.

Gallipagoes Islands, 259, 287.
Ganges *(ship)*, 317
Gardner's Island, 317.
Gasper's Island, 169.
Gelston, David, 300.
Gherritz, Captain, 304.
Gomez, Joseph Maria, 272.
Good News *(ship)*, 301.
Granger's Island, 319.
Great Ganges Island, 317.
Group Islands, 317.

Hall, —, 184, 189.
Havel, Matthew A., 269, 287.
Hersilia *(brig)*, 301, 305, 306.
Hermes *(ship)*, 318.
Hero *(sloop)*, 306, 307.
Holystone, 15.
Hood's Island, 83.
Hope *(ship)*, 239, 240.
Hunter, Captain, 318.
Hunting and fishing, 218.

Ice islands, 220.
Icebergs, 312.
Impressment of American seamen, 207.
Independence *(ship)*, 316, 318.

James Monroe *(sloop)*, 309.
Japan *(ship)*, 318.
Java Head, 201.
Jefferson's Island, 317.
John Adams *(frigate)*, 14.
John and Elizabeth *(ship)*, 49.

Kidd's Island, 320.
King George III Sound, 232, 243.

La Christiana Island, 87.
La Perouse, 106.
L'Uranie *(French corvette)*, 267.
Ladrone Islands, 179.
Ladrone pirates, 193-200.
Lancaster Reef, 317.
Letter of marque, 206.
Lightning strikes ship, 298.
Lincoln's Island, 318.
Lintin, China, 186.
Little Ganges Island, 317.
Loper's Island, 319.
Lord, —, 234, 239.

Macao roads, 225.
McClannon, Captain, 174.
Mackay, D., 298.
Mackay, Capt. Donald, 160, 168.
McKenzie, —, 185.
Maro *(ship)*, 319.
Marquesas Islands, 84-155.

334　INDEX

Massafuero Island, 63, 73, 267.
Melon, Capt. L., 237.
Miller, Captain, 1.
Missionary, Rescue of, 89.
Mitchell, Capt. J., 317.
Mitchell's Group, 316.
Morrell, Benjamin, 294.
Morse *(ship)*, 216, 219.
Mosey, Eliza, 237.
Mother of pearl, 329.
Motley, Captain, 42.

New Discovery Island, 318.
New Haven Packet, 300.
New Holland, 231, 242.
New London, Conn., 299.
New Nantucket, 319.
New South Wales, 233.
New York *(frigate)*, 14.
New York Island, 155, 247.
Nexsen, Elias, 2, 4, 13, 41, 155.
Nexsen Island, 155.
Nixon's Rock, 318.
Nuggoheeva Island, 108.
Nymphe *(frigate)*, 22.

Oeno Island, 317.
Oeno *(ship)*, 318.
Olive Branch *(whale ship)*, 63, 65.
Ontario *(ship)*, 40, 194.
Ontario *(sloop of war)*, 294.

Pacific *(schooner)*, 310.
Paddock, Capt. O., 63, 65, 73.
Palmer, Capt. N. B., 306, 307.
Palmer's Land, 258, 307, 309.
Palmyra's Island, 168.

Parker's Island, 318.
Parry, —, 185.
Patagonia, 53.
Patagonian giants, 66.
Pearl *(ship)*, 318.
Pellew, Sir Edward, 22, 25.
Pendleton, B., 281, 289.
Pendleton, Capt. Benjamin, 306, 307, 309.
Pendleton, Capt. Isaac, 230, 234, 239, 240.
Penguin, 56.
Penguin eggs, 56, 60, 63.
Percival, Capt. —, 248.
Pernambuco, 209.
Phillips, Captain, 318.
Pike's Island, 316.
Pirates, 193-200, 250-253.
Plasket, Captain, 318.
Portland *(ship)*, 14, 17, 24, 34.
Portsmouth, 310.
Potatoes, 301.
Potter's Island, 311.
Prince, —, 312.
Prince Edward's Islands, 246.

Rambler *(ship)*, 317.
Rambler's Reef, 316.
Reaper Island, 317.
Regulator *(vessel)*, 211, 215-217.
Rocky Island, 316.
Robinson, Capt. Thomas, 14, 19, 34, 39.
Romulus *(frigate)*, 46.
Rossiter, Asa, 1.
Russian vessels, 307.

St. Helena, 202.
St. Mary's, Chili, 223, 288.

INDEX 335

Sandal Wood, 192, 226, 239, 322.
Saypan, 224.
Saxenburgh Island, 211.
Sea elephants, 213, 255, 257.
Sea Fox *(ship)*, 295, 300.
Sea hens, 213.
Sea leopard, 257, 258, 310.
Sea lions, 9, 58, 258-260.
Sea tiger, 312.
Seal Island, 231, 243.
Seal skins, 79, 218.
Seals, 8, 79, 217, 232, 248, 260-264, 287, 295, 304.
Scurvy, 204.
Shark (shovel-nose), 7.
Sheffield, —, 212.
Sheffield, Capt. James P., 301, 306.
Shipwreck, 171.
Sidney, 233.
Smut-face Island, 318.
Snell, Stagg & Co., 33.
South Georgia, 213, 215, 222, 301, 310.
South Shetlands, 305-307.
Starbuck's Group, 319.
Staten Island, 302.
Staten Island, N. Y., 208.
Steele, Capt. R., 4.
Stonington, Conn., 294, 306.
Swain, —, 171, 183.
Swain, Captain, 316.
Swain's Island, 316.

Tea, cargo of, damaged, 205.
Terra del Fuego, 223.
Thompson, C., 229.

Tinian Island, 170-178.
Tongataboo, 234, 240.
Tracy's Island, 319.
Transit *(ship)*, 319.
Tristian de Cunha, 211.
Tuck's Island, 316.
Tussock grass, 11.

Union *(brig)*, 230, 234, 239, 242.
Unknown Island, 317.
Unknown Reef, 318.

Valetta Island, 319.
Valetta *(ship)*, 319.
Vancouver, —, 230.
Van Diemen's Land, 233, 240.
Volcanic eruption, 5.
Volcano, 311.
Volunteer *(ship)*, 265, 293.

Wampoa, 188.
Washington Islands, 99, 161.
Weeks, Captain, 317.
Whale Rock, 319.
Whales, 224, 296.
Whetten, Capt. John, 40.
Whetton, William, 4, 13.
Willey's Island, 312.
William *(brig)*, 304.
Williams, Capt. E., 306.
Willis Island, 215.
Wilson, Captain, 89.
Worth, Capt. J. B., 318.
Worth, Capt. William, 317.
Worthington, —, 293.
Worth's Island, 316.
Wright, —, 15.
Wright, D., 230, 234.

A CATALOG OF SELECTED
DOVER BOOKS
IN ALL FIELDS OF INTEREST

A CATALOG OF SELECTED DOVER BOOKS IN ALL FIELDS OF INTEREST

DRAWINGS OF REMBRANDT, edited by Seymour Slive. Updated Lippmann, Hofstede de Groot edition, with definitive scholarly apparatus. All portraits, biblical sketches, landscapes, nudes. Oriental figures, classical studies, together with selection of work by followers. 550 illustrations. Total of 630pp. 9⅛ × 12¼.
21485-0, 21486-9 Pa., Two-vol. set $25.00

GHOST AND HORROR STORIES OF AMBROSE BIERCE, Ambrose Bierce. 24 tales vividly imagined, strangely prophetic, and decades ahead of their time in technical skill: "The Damned Thing," "An Inhabitant of Carcosa," "The Eyes of the Panther," "Moxon's Master," and 20 more. 199pp. 5⅜ × 8½. 20767-6 Pa. $3.95

ETHICAL WRITINGS OF MAIMONIDES, Maimonides. Most significant ethical works of great medieval sage, newly translated for utmost precision, readability. Laws Concerning Character Traits, Eight Chapters, more. 192pp. 5⅜ × 8½.
24522-5 Pa. $4.50

THE EXPLORATION OF THE COLORADO RIVER AND ITS CANYONS, J. W. Powell. Full text of Powell's 1,000-mile expedition down the fabled Colorado in 1869. Superb account of terrain, geology, vegetation, Indians, famine, mutiny, treacherous rapids, mighty canyons, during exploration of last unknown part of continental U.S. 400pp. 5⅜ × 8½. 20094-9 Pa. $6.95

HISTORY OF PHILOSOPHY, Julián Marías. Clearest one-volume history on the market. Every major philosopher and dozens of others, to Existentialism and later. 505pp. 5⅜ × 8½. 21739-6 Pa. $8.50

ALL ABOUT LIGHTNING, Martin A. Uman. Highly readable non-technical survey of nature and causes of lightning, thunderstorms, ball lightning, St. Elmo's Fire, much more. Illustrated. 192pp. 5⅜ × 8½. 25237-X Pa. $5.95

SAILING ALONE AROUND THE WORLD, Captain Joshua Slocum. First man to sail around the world, alone, in small boat. One of great feats of seamanship told in delightful manner. 67 illustrations. 294pp. 5⅜ × 8½. 20326-3 Pa. $4.95

LETTERS AND NOTES ON THE MANNERS, CUSTOMS AND CONDITIONS OF THE NORTH AMERICAN INDIANS, George Catlin. Classic account of life among Plains Indians: ceremonies, hunt, warfare, etc. 312 plates. 572pp. of text. 6⅛ × 9¼. 22118-0, 22119-9 Pa. Two-vol. set $15.90

ALASKA: The Harriman Expedition, 1899, John Burroughs, John Muir, et al. Informative, engrossing accounts of two-month, 9,000-mile expedition. Native peoples, wildlife, forests, geography, salmon industry, glaciers, more. Profusely illustrated. 240 black-and-white line drawings. 124 black-and-white photographs. 3 maps. Index. 576pp. 5⅜ × 8½. 25109-8 Pa. $11.95

CATALOG OF DOVER BOOKS

THE BOOK OF BEASTS: Being a Translation from a Latin Bestiary of the Twelfth Century, T. H. White. Wonderful catalog real and fanciful beasts: manticore, griffin, phoenix, amphivius, jaculus, many more. White's witty erudite commentary on scientific, historical aspects. Fascinating glimpse of medieval mind. Illustrated. 296pp. 5⅜ × 8¼. (Available in U.S. only) 24609-4 Pa. $5.95

FRANK LLOYD WRIGHT: ARCHITECTURE AND NATURE With 160 Illustrations, Donald Hoffmann. Profusely illustrated study of influence of nature—especially prairie—on Wright's designs for Fallingwater, Robie House, Guggenheim Museum, other masterpieces. 96pp. 9¼ × 10¾. 25098-9 Pa. $7.95

FRANK LLOYD WRIGHT'S FALLINGWATER, Donald Hoffmann. Wright's famous waterfall house: planning and construction of organic idea. History of site, owners, Wright's personal involvement. Photographs of various stages of building. Preface by Edgar Kaufmann, Jr. 100 illustrations. 112pp. 9¼ × 10. 23671-4 Pa. $7.95

YEARS WITH FRANK LLOYD WRIGHT: Apprentice to Genius, Edgar Tafel. Insightful memoir by a former apprentice presents a revealing portrait of Wright the man, the inspired teacher, the greatest American architect. 372 black-and-white illustrations. Preface. Index. vi + 228pp. 8¼ × 11. 24801-1 Pa. $9.95

THE STORY OF KING ARTHUR AND HIS KNIGHTS, Howard Pyle. Enchanting version of King Arthur fable has delighted generations with imaginative narratives of exciting adventures and unforgettable illustrations by the author. 41 illustrations. xviii + 313pp. 6⅛ × 9¼. 21445-1 Pa. $6.50

THE GODS OF THE EGYPTIANS, E. A. Wallis Budge. Thorough coverage of numerous gods of ancient Egypt by foremost Egyptologist. Information on evolution of cults, rites and gods; the cult of Osiris; the Book of the Dead and its rites; the sacred animals and birds; Heaven and Hell; and more. 956pp. 6⅛ × 9¼. 22055-9, 22056-7 Pa., Two-vol. set $20.00

A THEOLOGICO-POLITICAL TREATISE, Benedict Spinoza. Also contains unfinished *Political Treatise*. Great classic on religious liberty, theory of government on common consent. R. Elwes translation. Total of 421pp. 5⅜ × 8½. 20249-6 Pa. $6.95

INCIDENTS OF TRAVEL IN CENTRAL AMERICA, CHIAPAS, AND YUCATAN, John L. Stephens. Almost single-handed discovery of Maya culture; exploration of ruined cities, monuments, temples; customs of Indians. 115 drawings. 892pp. 5⅜ × 8½. 22404-X, 22405-8 Pa., Two-vol. set $15.90

LOS CAPRICHOS, Francisco Goya. 80 plates of wild, grotesque monsters and caricatures. Prado manuscript included. 183pp. 6⅞ × 9⅞. 22384-1 Pa. $4.95

AUTOBIOGRAPHY: The Story of My Experiments with Truth, Mohandas K. Gandhi. Not hagiography, but Gandhi in his own words. Boyhood, legal studies, purification, the growth of the Satyagraha (nonviolent protest) movement. Critical, inspiring work of the man who freed India. 480pp. 5⅜ × 8½. (Available in U.S. only) 24593-4 Pa. $6.95

CATALOG OF DOVER BOOKS

ILLUSTRATED DICTIONARY OF HISTORIC ARCHITECTURE, edited by Cyril M. Harris. Extraordinary compendium of clear, concise definitions for over 5,000 important architectural terms complemented by over 2,000 line drawings. Covers full spectrum of architecture from ancient ruins to 20th-century Modernism. Preface. 592pp. 7½ × 9⅜. 24444-X Pa. $14.95

THE NIGHT BEFORE CHRISTMAS, Clement Moore. Full text, and woodcuts from original 1848 book. Also critical, historical material. 19 illustrations. 40pp. 4⅝ × 6. 22797-9 Pa. $2.25

THE LESSON OF JAPANESE ARCHITECTURE: 165 Photographs, Jiro Harada. Memorable gallery of 165 photographs taken in the 1930's of exquisite Japanese homes of the well-to-do and historic buildings. 13 line diagrams. 192pp. 8⅜ × 11¼. 24778-3 Pa. $8.95

THE AUTOBIOGRAPHY OF CHARLES DARWIN AND SELECTED LETTERS, edited by Francis Darwin. The fascinating life of eccentric genius composed of an intimate memoir by Darwin (intended for his children); commentary by his son, Francis; hundreds of fragments from notebooks, journals, papers; and letters to and from Lyell, Hooker, Huxley, Wallace and Henslow. xi + 365pp. 5⅜ × 8. 20479-0 Pa. $6.95

WONDERS OF THE SKY: Observing Rainbows, Comets, Eclipses, the Stars and Other Phenomena, Fred Schaaf. Charming, easy-to-read poetic guide to all manner of celestial events visible to the naked eye. Mock suns, glories, Belt of Venus, more. Illustrated. 299pp. 5¼ × 8¼. 24402-4 Pa. $7.95

BURNHAM'S CELESTIAL HANDBOOK, Robert Burnham, Jr. Thorough guide to the stars beyond our solar system. Exhaustive treatment. Alphabetical by constellation: Andromeda to Cetus in Vol. 1; Chamaeleon to Orion in Vol. 2; and Pavo to Vulpecula in Vol. 3. Hundreds of illustrations. Index in Vol. 3. 2,000pp. 6⅛ × 9¼. 23567-X, 23568-8, 23673-0 Pa., Three-vol. set $38.85

STAR NAMES: Their Lore and Meaning, Richard Hinckley Allen. Fascinating history of names various cultures have given to constellations and literary and folkloristic uses that have been made of stars. Indexes to subjects. Arabic and Greek names. Biblical references. Bibliography. 563pp. 5⅜ × 8½. 21079-0 Pa. $7.95

THIRTY YEARS THAT SHOOK PHYSICS: The Story of Quantum Theory, George Gamow. Lucid, accessible introduction to influential theory of energy and matter. Careful explanations of Dirac's anti-particles, Bohr's model of the atom, much more. 12 plates. Numerous drawings. 240pp. 5⅜ × 8½. 24895-X Pa. $4.95

CHINESE DOMESTIC FURNITURE IN PHOTOGRAPHS AND MEASURED DRAWINGS, Gustav Ecke. A rare volume, now affordably priced for antique collectors, furniture buffs and art historians. Detailed review of styles ranging from early Shang to late Ming. Unabridged republication. 161 black-and-white drawings, photos. Total of 224pp. 8⅜ × 11¼. (Available in U.S. only) 25171-3 Pa. $12.95

VINCENT VAN GOGH: A Biography, Julius Meier-Graefe. Dynamic, penetrating study of artist's life, relationship with brother, Theo, painting techniques, travels, more. Readable, engrossing. 160pp. 5⅜ × 8½. (Available in U.S. only) 25253-1 Pa. $3.95

HOW TO WRITE, Gertrude Stein. Gertrude Stein claimed anyone could understand her unconventional writing—here are clues to help. Fascinating improvisations, language experiments, explanations illuminate Stein's craft and the art of writing. Total of 414pp. 4⅝ × 6⅝. 23144-5 Pa. $5.95

ADVENTURES AT SEA IN THE GREAT AGE OF SAIL: Five Firsthand Narratives, edited by Elliot Snow. Rare true accounts of exploration, whaling, shipwreck, fierce natives, trade, shipboard life, more. 33 illustrations. Introduction. 353pp. 5⅜ × 8½. 25177-2 Pa. $7.95

THE HERBAL OR GENERAL HISTORY OF PLANTS, John Gerard. Classic descriptions of about 2,850 plants—with over 2,700 illustrations—includes Latin and English names, physical descriptions, varieties, time and place of growth, more. 2,706 illustrations. xlv + 1,678pp. 8½ × 12¼. 23147-X Cloth. $75.00

DOROTHY AND THE WIZARD IN OZ, L. Frank Baum. Dorothy and the Wizard visit the center of the Earth, where people are vegetables, glass houses grow and Oz characters reappear. Classic sequel to *Wizard of Oz*. 256pp. 5⅜ × 8.
24714-7 Pa. $4.95

SONGS OF EXPERIENCE: Facsimile Reproduction with 26 Plates in Full Color, William Blake. This facsimile of Blake's original "Illuminated Book" reproduces 26 full-color plates from a rare 1826 edition. Includes "The Tyger," "London," "Holy Thursday," and other immortal poems. 26 color plates. Printed text of poems. 48pp. 5¼ × 7. 24636-1 Pa. $3.50

SONGS OF INNOCENCE, William Blake. The first and most popular of Blake's famous "Illuminated Books," in a facsimile edition reproducing all 31 brightly colored plates. Additional printed text of each poem. 64pp. 5¼ × 7.
22764-2 Pa. $3.50

PRECIOUS STONES, Max Bauer. Classic, thorough study of diamonds, rubies, emeralds, garnets, etc.: physical character, occurrence, properties, use, similar topics. 20 plates, 8 in color. 94 figures. 659pp. 6⅛ × 9¼.
21910-0, 21911-9 Pa., Two-vol. set $15.90

ENCYCLOPEDIA OF VICTORIAN NEEDLEWORK, S. F. A. Caulfeild and Blanche Saward. Full, precise descriptions of stitches, techniques for dozens of needlecrafts—most exhaustive reference of its kind. Over 800 figures. Total of 679pp. 8⅛ × 11. Two volumes. Vol. 1 22800-2 Pa. $11.95
Vol. 2 22801-0 Pa. $11.95

THE MARVELOUS LAND OF OZ, L. Frank Baum. Second Oz book, the Scarecrow and Tin Woodman are back with hero named Tip, Oz magic. 136 illustrations. 287pp. 5⅜ × 8½. 20692-0 Pa. $5.95

WILD FOWL DECOYS, Joel Barber. Basic book on the subject, by foremost authority and collector. Reveals history of decoy making and rigging, place in American culture, different kinds of decoys, how to make them, and how to use them. 140 plates. 156pp. 7⅞ × 10¾. 20011-6 Pa. $8.95

HISTORY OF LACE, Mrs. Bury Palliser. Definitive, profusely illustrated chronicle of lace from earliest times to late 19th century. Laces of Italy, Greece, England, France, Belgium, etc. Landmark of needlework scholarship. 266 illustrations. 672pp. 6⅛ × 9¼. 24742-2 Pa. $14.95

CATALOG OF DOVER BOOKS

ILLUSTRATED GUIDE TO SHAKER FURNITURE, Robert Meader. All furniture and appurtenances, with much on unknown local styles. 235 photos. 146pp. 9 × 12. 22819-3 Pa. $7.95

WHALE SHIPS AND WHALING: A Pictorial Survey, George Francis Dow. Over 200 vintage engravings, drawings, photographs of barks, brigs, cutters, other vessels. Also harpoons, lances, whaling guns, many other artifacts. Comprehensive text by foremost authority. 207 black-and-white illustrations. 288pp. 6 × 9. 24808-9 Pa. $8.95

THE BERTRAMS, Anthony Trollope. Powerful portrayal of blind self-will and thwarted ambition includes one of Trollope's most heartrending love stories. 497pp. 5⅜ × 8½. 25119-5 Pa. $8.95

ADVENTURES WITH A HAND LENS, Richard Headstrom. Clearly written guide to observing and studying flowers and grasses, fish scales, moth and insect wings, egg cases, buds, feathers, seeds, leaf scars, moss, molds, ferns, common crystals, etc.—all with an ordinary, inexpensive magnifying glass. 209 exact line drawings aid in your discoveries. 220pp. 5⅜ × 8½. 23330-8 Pa. $3.95

RODIN ON ART AND ARTISTS, Auguste Rodin. Great sculptor's candid, wide-ranging comments on meaning of art; great artists; relation of sculpture to poetry, painting, music; philosophy of life, more. 76 superb black-and-white illustrations of Rodin's sculpture, drawings and prints. 119pp. 8⅜ × 11¼. 24487-3 Pa. $6.95

FIFTY CLASSIC FRENCH FILMS, 1912–1982: A Pictorial Record, Anthony Slide. Memorable stills from Grand Illusion, Beauty and the Beast, Hiroshima, Mon Amour, many more. Credits, plot synopses, reviews, etc. 160pp. 8¼ × 11. 25256-6 Pa. $11.95

THE PRINCIPLES OF PSYCHOLOGY, William James. Famous long course complete, unabridged. Stream of thought, time perception, memory, experimental methods; great work decades ahead of its time. 94 figures. 1,391pp. 5⅜ × 8½. 20381-6, 20382-4 Pa., Two-vol. set $19.90

BODIES IN A BOOKSHOP, R. T. Campbell. Challenging mystery of blackmail and murder with ingenious plot and superbly drawn characters. In the best tradition of British suspense fiction. 192pp. 5⅜ × 8½. 24720-1 Pa. $3.95

CALLAS: PORTRAIT OF A PRIMA DONNA, George Jellinek. Renowned commentator on the musical scene chronicles incredible career and life of the most controversial, fascinating, influential operatic personality of our time. 64 black-and-white photographs. 416pp. 5⅜ × 8¼. 25047-4 Pa. $7.95

GEOMETRY, RELATIVITY AND THE FOURTH DIMENSION, Rudolph Rucker. Exposition of fourth dimension, concepts of relativity as Flatland characters continue adventures. Popular, easily followed yet accurate, profound. 141 illustrations. 133pp. 5⅜ × 8½. 23400-2 Pa. $3.95

HOUSEHOLD STORIES BY THE BROTHERS GRIMM, with pictures by Walter Crane. 53 classic stories—Rumpelstiltskin, Rapunzel, Hansel and Gretel, the Fisherman and his Wife, Snow White, Tom Thumb, Sleeping Beauty, Cinderella, and so much more—lavishly illustrated with original 19th century drawings. 114 illustrations. x + 269pp. 5⅜ × 8½. 21080-4 Pa. $4.50

SUNDIALS, Albert Waugh. Far and away the best, most thorough coverage of ideas, mathematics concerned, types, construction, adjusting anywhere. Over 100 illustrations. 230pp. 5⅜ × 8½. 22947-5 Pa. $4.50

PICTURE HISTORY OF THE NORMANDIE: With 190 Illustrations, Frank O. Braynard. Full story of legendary French ocean liner: Art Deco interiors, design innovations, furnishings, celebrities, maiden voyage, tragic fire, much more. Extensive text. 144pp. 8⅞ × 11¾. 25257-4 Pa. $9.95

THE FIRST AMERICAN COOKBOOK: A Facsimile of "American Cookery," 1796, Amelia Simmons. Facsimile of the first American-written cookbook published in the United States contains authentic recipes for colonial favorites—pumpkin pudding, winter squash pudding, spruce beer, Indian slapjacks, and more. Introductory Essay and Glossary of colonial cooking terms. 80pp. 5⅜ × 8½. 24710-4 Pa. $3.50

101 PUZZLES IN THOUGHT AND LOGIC, C. R. Wylie, Jr. Solve murders and robberies, find out which fishermen are liars, how a blind man could possibly identify a color—purely by your own reasoning! 107pp. 5⅜ × 8½. 20367-0 Pa. $2.50

THE BOOK OF WORLD-FAMOUS MUSIC—CLASSICAL, POPULAR AND FOLK, James J. Fuld. Revised and enlarged republication of landmark work in musico-bibliography. Full information about nearly 1,000 songs and compositions including first lines of music and lyrics. New supplement. Index. 800pp. 5⅜ × 8¼. 24857-7 Pa. $14.95

ANTHROPOLOGY AND MODERN LIFE, Franz Boas. Great anthropologist's classic treatise on race and culture. Introduction by Ruth Bunzel. Only inexpensive paperback edition. 255pp. 5⅜ × 8½. 25245-0 Pa. $5.95

THE TALE OF PETER RABBIT, Beatrix Potter. The inimitable Peter's terrifying adventure in Mr. McGregor's garden, with all 27 wonderful, full-color Potter illustrations. 55pp. 4¼ × 5½. (Available in U.S. only) 22827-4 Pa. $1.75

THREE PROPHETIC SCIENCE FICTION NOVELS, H. G. Wells. *When the Sleeper Wakes, A Story of the Days to Come* and *The Time Machine* (full version). 335pp. 5⅜ × 8½. (Available in U.S. only) 20605-X Pa. $5.95

APICIUS COOKERY AND DINING IN IMPERIAL ROME, edited and translated by Joseph Dommers Vehling. Oldest known cookbook in existence offers readers a clear picture of what foods Romans ate, how they prepared them, etc. 49 illustrations. 301pp. 6⅛ × 9¼. 23563-7 Pa. $6.50

SHAKESPEARE LEXICON AND QUOTATION DICTIONARY, Alexander Schmidt. Full definitions, locations, shades of meaning of every word in plays and poems. More than 50,000 exact quotations. 1,485pp. 6½ × 9¼. 22726-X, 22727-8 Pa., Two-vol. set $27.90

THE WORLD'S GREAT SPEECHES, edited by Lewis Copeland and Lawrence W. Lamm. Vast collection of 278 speeches from Greeks to 1970. Powerful and effective models; unique look at history. 842pp. 5⅜ × 8½. 20468-5 Pa. $11.95

CATALOG OF DOVER BOOKS

THE BLUE FAIRY BOOK, Andrew Lang. The first, most famous collection, with many familiar tales: Little Red Riding Hood, Aladdin and the Wonderful Lamp, Puss in Boots, Sleeping Beauty, Hansel and Gretel, Rumpelstiltskin; 37 in all. 138 illustrations. 390pp. 5⅜ × 8½. 21437-0 Pa. $5.95

THE STORY OF THE CHAMPIONS OF THE ROUND TABLE, Howard Pyle. Sir Launcelot, Sir Tristram and Sir Percival in spirited adventures of love and triumph retold in Pyle's inimitable style. 50 drawings, 31 full-page. xviii + 329pp. 6½ × 9¼. 21883-X Pa. $6.95

AUDUBON AND HIS JOURNALS, Maria Audubon. Unmatched two-volume portrait of the great artist, naturalist and author contains his journals, an excellent biography by his granddaughter, expert annotations by the noted ornithologist, Dr. Elliott Coues, and 37 superb illustrations. Total of 1,200pp. 5⅜ × 8.
Vol. I 25143-8 Pa. $8.95
Vol. II 25144-6 Pa. $8.95

GREAT DINOSAUR HUNTERS AND THEIR DISCOVERIES, Edwin H. Colbert. Fascinating, lavishly illustrated chronicle of dinosaur research, 1820's to 1960. Achievements of Cope, Marsh, Brown, Buckland, Mantell, Huxley, many others. 384pp. 5¼ × 8¼. 24701-5 Pa. $6.95

THE TASTEMAKERS, Russell Lynes. Informal, illustrated social history of American taste 1850's-1950's. First popularized categories Highbrow, Lowbrow, Middlebrow. 129 illustrations. New (1979) afterword. 384pp. 6 × 9.
23993-4 Pa. $6.95

DOUBLE CROSS PURPOSES, Ronald A. Knox. A treasure hunt in the Scottish Highlands, an old map, unidentified corpse, surprise discoveries keep reader guessing in this cleverly intricate tale of financial skullduggery. 2 black-and-white maps. 320pp. 5⅜ × 8½. (Available in U.S. only) 25032-6 Pa. $5.95

AUTHENTIC VICTORIAN DECORATION AND ORNAMENTATION IN FULL COLOR: 46 Plates from "Studies in Design," Christopher Dresser. Superb full-color lithographs reproduced from rare original portfolio of a major Victorian designer. 48pp. 9¼ × 12¼. 25083-0 Pa. $7.95

PRIMITIVE ART, Franz Boas. Remains the best text ever prepared on subject, thoroughly discussing Indian, African, Asian, Australian, and, especially, Northern American primitive art. Over 950 illustrations show ceramics, masks, totem poles, weapons, textiles, paintings, much more. 376pp. 5⅜ × 8. 20025-6 Pa. $6.95

SIDELIGHTS ON RELATIVITY, Albert Einstein. Unabridged republication of two lectures delivered by the great physicist in 1920-21. *Ether and Relativity* and *Geometry and Experience*. Elegant ideas in non-mathematical form, accessible to intelligent layman. vi + 56pp. 5⅜ × 8½. 24511-X Pa. $2.95

THE WIT AND HUMOR OF OSCAR WILDE, edited by Alvin Redman. More than 1,000 ripostes, paradoxes, wisecracks: Work is the curse of the drinking classes, I can resist everything except temptation, etc. 258pp. 5⅜ × 8½. 20602-5 Pa. $4.50

ADVENTURES WITH A MICROSCOPE, Richard Headstrom. 59 adventures with clothing fibers, protozoa, ferns and lichens, roots and leaves, much more. 142 illustrations. 232pp. 5⅜ × 8½. 23471-1 Pa. $3.95

CATALOG OF DOVER BOOKS

PLANTS OF THE BIBLE, Harold N. Moldenke and Alma L. Moldenke. Standard reference to all 230 plants mentioned in Scriptures. Latin name, biblical reference, uses, modern identity, much more. Unsurpassed encyclopedic resource for scholars, botanists, nature lovers, students of Bible. Bibliography. Indexes. 123 black-and-white illustrations. 384pp. 6 × 9. 25069-5 Pa. $8.95

FAMOUS AMERICAN WOMEN: A Biographical Dictionary from Colonial Times to the Present, Robert McHenry, ed. From Pocahontas to Rosa Parks, 1,035 distinguished American women documented in separate biographical entries. Accurate, up-to-date data, numerous categories, spans 400 years. Indices. 493pp. 6½ × 9¼. 24523-3 Pa. $9.95

THE FABULOUS INTERIORS OF THE GREAT OCEAN LINERS IN HISTORIC PHOTOGRAPHS, William H. Miller, Jr. Some 200 superb photographs capture exquisite interiors of world's great "floating palaces"—1890's to 1980's: *Titanic, Ile de France, Queen Elizabeth, United States, Europa*, more. Approx. 200 black-and-white photographs. Captions. Text. Introduction. 160pp. 8⅜ × 11¼. 24756-2 Pa. $9.95

THE GREAT LUXURY LINERS, 1927-1954: A Photographic Record, William H. Miller, Jr. Nostalgic tribute to heyday of ocean liners. 186 photos of Ile de France, Normandie, Leviathan, Queen Elizabeth, United States, many others. Interior and exterior views. Introduction. Captions. 160pp. 9 × 12. 24056-8 Pa. $9.95

A NATURAL HISTORY OF THE DUCKS, John Charles Phillips. Great landmark of ornithology offers complete detailed coverage of nearly 200 species and subspecies of ducks: gadwall, sheldrake, merganser, pintail, many more. 74 full-color plates, 102 black-and-white. Bibliography. Total of 1,920pp. 8⅜ × 11¼. 25141-1, 25142-X Cloth. Two-vol. set $100.00

THE SEAWEED HANDBOOK: An Illustrated Guide to Seaweeds from North Carolina to Canada, Thomas F. Lee. Concise reference covers 78 species. Scientific and common names, habitat, distribution, more. Finding keys for easy identification. 224pp. 5⅜ × 8½. 25215-9 Pa. $5.95

THE TEN BOOKS OF ARCHITECTURE: The 1755 Leoni Edition, Leon Battista Alberti. Rare classic helped introduce the glories of ancient architecture to the Renaissance. 68 black-and-white plates. 336pp. 8⅜ × 11¼. 25239-6 Pa. $14.95

MISS MACKENZIE, Anthony Trollope. Minor masterpieces by Victorian master unmasks many truths about life in 19th-century England. First inexpensive edition in years. 392pp. 5⅜ × 8½. 25201-9 Pa. $7.95

THE RIME OF THE ANCIENT MARINER, Gustave Doré, Samuel Taylor Coleridge. Dramatic engravings considered by many to be his greatest work. The terrifying space of the open sea, the storms and whirlpools of an unknown ocean, the ice of Antarctica, more—all rendered in a powerful, chilling manner. Full text. 38 plates. 77pp. 9¼ × 12. 22305-1 Pa. $4.95

THE EXPEDITIONS OF ZEBULON MONTGOMERY PIKE, Zebulon Montgomery Pike. Fascinating first-hand accounts (1805-6) of exploration of Mississippi River, Indian wars, capture by Spanish dragoons, much more. 1,088pp. 5⅜ × 8½. 25254-X, 25255-8 Pa. Two-vol. set $23.90

CATALOG OF DOVER BOOKS

A CONCISE HISTORY OF PHOTOGRAPHY: Third Revised Edition, Helmut Gernsheim. Best one-volume history—camera obscura, photochemistry, daguerreotypes, evolution of cameras, film, more. Also artistic aspects—landscape, portraits, fine art, etc. 281 black-and-white photographs. 26 in color. 176pp. 8⅜ × 11¼. 25128-4 Pa. $12.95

THE DORÉ BIBLE ILLUSTRATIONS, Gustave Doré. 241 detailed plates from the Bible: the Creation scenes, Adam and Eve, Flood, Babylon, battle sequences, life of Jesus, etc. Each plate is accompanied by the verses from the King James version of the Bible. 241pp. 9 × 12. 23004-X Pa. $8.95

HUGGER-MUGGER IN THE LOUVRE, Elliot Paul. Second Homer Evans mystery-comedy. Theft at the Louvre involves sleuth in hilarious, madcap caper. "A knockout."—Books. 336pp. 5⅜ × 8½. 25185-3 Pa. $5.95

FLATLAND, E. A. Abbott. Intriguing and enormously popular science-fiction classic explores the complexities of trying to survive as a two-dimensional being in a three-dimensional world. Amusingly illustrated by the author. 16 illustrations. 103pp. 5⅜ × 8½. 20001-9 Pa. $2.25

THE HISTORY OF THE LEWIS AND CLARK EXPEDITION, Meriwether Lewis and William Clark, edited by Elliott Coues. Classic edition of Lewis and Clark's day-by-day journals that later became the basis for U.S. claims to Oregon and the West. Accurate and invaluable geographical, botanical, biological, meteorological and anthropological material. Total of 1,508pp. 5⅜ × 8½.
21268-8, 21269-6, 21270-X Pa. Three-vol. set $25.50

LANGUAGE, TRUTH AND LOGIC, Alfred J. Ayer. Famous, clear introduction to Vienna, Cambridge schools of Logical Positivism. Role of philosophy, elimination of metaphysics, nature of analysis, etc. 160pp. 5⅜ × 8½. (Available in U.S. and Canada only) 20010-8 Pa. $2.95

MATHEMATICS FOR THE NONMATHEMATICIAN, Morris Kline. Detailed, college-level treatment of mathematics in cultural and historical context, with numerous exercises. For liberal arts students. Preface. Recommended Reading Lists. Tables. Index. Numerous black-and-white figures. xvi + 641pp. 5⅜ × 8½.
24823-2 Pa. $11.95

28 SCIENCE FICTION STORIES, H. G. Wells. Novels, *Star Begotten* and *Men Like Gods*, plus 26 short stories: "Empire of the Ants," "A Story of the Stone Age," "The Stolen Bacillus," "In the Abyss," etc. 915pp. 5⅜ × 8½. (Available in U.S. only)
20265-8 Cloth. $10.95

HANDBOOK OF PICTORIAL SYMBOLS, Rudolph Modley. 3,250 signs and symbols, many systems in full; official or heavy commercial use. Arranged by subject. Most in Pictorial Archive series. 143pp. 8⅜ × 11. 23357-X Pa. $5.95

INCIDENTS OF TRAVEL IN YUCATAN, John L. Stephens. Classic (1843) exploration of jungles of Yucatan, looking for evidences of Maya civilization. Travel adventures, Mexican and Indian culture, etc. Total of 669pp. 5⅜ × 8½.
20926-1, 20927-X Pa., Two-vol. set $9.90

CATALOG OF DOVER BOOKS

DEGAS: An Intimate Portrait, Ambroise Vollard. Charming, anecdotal memoir by famous art dealer of one of the greatest 19th-century French painters. 14 black-and-white illustrations. Introduction by Harold L. Van Doren. 96pp. 5⅜ × 8½.
25131-4 Pa. $3.95

PERSONAL NARRATIVE OF A PILGRIMAGE TO ALMANDINAH AND MECCAH, Richard Burton. Great travel classic by remarkably colorful personality. Burton, disguised as a Moroccan, visited sacred shrines of Islam, narrowly escaping death. 47 illustrations. 959pp. 5⅜ × 8½. 21217-3, 21218-1 Pa., Two-vol. set $19.90

PHRASE AND WORD ORIGINS, A. H. Holt. Entertaining, reliable, modern study of more than 1,200 colorful words, phrases, origins and histories. Much unexpected information. 254pp. 5⅜ × 8½.
20758-7 Pa. $4.95

THE RED THUMB MARK, R. Austin Freeman. In this first Dr. Thorndyke case, the great scientific detective draws fascinating conclusions from the nature of a single fingerprint. Exciting story, authentic science. 320pp. 5⅜ × 8½. (Available in U.S. only)
25210-8 Pa. $5.95

AN EGYPTIAN HIEROGLYPHIC DICTIONARY, E. A. Wallis Budge. Monumental work containing about 25,000 words or terms that occur in texts ranging from 3000 B.C. to 600 A.D. Each entry consists of a transliteration of the word, the word in hieroglyphs, and the meaning in English. 1,314pp. 6⅜ × 10.
23615-3, 23616-1 Pa., Two-vol. set $27.90

THE COMPLEAT STRATEGYST: Being a Primer on the Theory of Games of Strategy, J. D. Williams. Highly entertaining classic describes, with many illustrated examples, how to select best strategies in conflict situations. Prefaces. Appendices. xvi + 268pp. 5⅜ × 8½.
25101-2 Pa. $5.95

THE ROAD TO OZ, L. Frank Baum. Dorothy meets the Shaggy Man, little Button-Bright and the Rainbow's beautiful daughter in this delightful trip to the magical Land of Oz. 272pp. 5⅜ × 8.
25208-6 Pa. $4.95

POINT AND LINE TO PLANE, Wassily Kandinsky. Seminal exposition of role of point, line, other elements in non-objective painting. Essential to understanding 20th-century art. 127 illustrations. 192pp. 6½ × 9¼.
23808-3 Pa. $4.50

LADY ANNA, Anthony Trollope. Moving chronicle of Countess Lovel's bitter struggle to win for herself and daughter Anna their rightful rank and fortune—perhaps at cost of sanity itself. 384pp. 5⅜ × 8½.
24669-8 Pa. $6.95

EGYPTIAN MAGIC, E. A. Wallis Budge. Sums up all that is known about magic in Ancient Egypt: the role of magic in controlling the gods, powerful amulets that warded off evil spirits, scarabs of immortality, use of wax images, formulas and spells, the secret name, much more. 253pp. 5⅜ × 8½.
22681-6 Pa. $4.00

THE DANCE OF SIVA, Ananda Coomaraswamy. Preeminent authority unfolds the vast metaphysic of India: the revelation of her art, conception of the universe, social organization, etc. 27 reproductions of art masterpieces. 192pp. 5⅜ × 8½.
24817-8 Pa. $5.95

CHRISTMAS CUSTOMS AND TRADITIONS, Clement A. Miles. Origin, evolution, significance of religious, secular practices. Caroling, gifts, yule logs, much more. Full, scholarly yet fascinating; non-sectarian. 400pp. 5⅜ × 8½.
23354-5 Pa. $6.50

THE HUMAN FIGURE IN MOTION, Eadweard Muybridge. More than 4,500 stopped-action photos, in action series, showing undraped men, women, children jumping, lying down, throwing, sitting, wrestling, carrying, etc. 390pp. 7⅞ × 10⅝.
20204-6 Cloth. $21.95

THE MAN WHO WAS THURSDAY, Gilbert Keith Chesterton. Witty, fast-paced novel about a club of anarchists in turn-of-the-century London. Brilliant social, religious, philosophical speculations. 128pp. 5⅜ × 8½. 25121-7 Pa. $3.95

A CEZANNE SKETCHBOOK: Figures, Portraits, Landscapes and Still Lifes, Paul Cezanne. Great artist experiments with tonal effects, light, mass, other qualities in over 100 drawings. A revealing view of developing master painter, precursor of Cubism. 102 black-and-white illustrations. 144pp. 8¾ × 6⅜. 24790-2 Pa. $5.95

AN ENCYCLOPEDIA OF BATTLES: Accounts of Over 1,560 Battles from 1479 B.C. to the Present, David Eggenberger. Presents essential details of every major battle in recorded history, from the first battle of Megiddo in 1479 B.C. to Grenada in 1984. List of Battle Maps. New Appendix covering the years 1967–1984. Index. 99 illustrations. 544pp. 6½ × 9¼. 24913-1 Pa. $14.95

AN ETYMOLOGICAL DICTIONARY OF MODERN ENGLISH, Ernest Weekley. Richest, fullest work, by foremost British lexicographer. Detailed word histories. Inexhaustible. Total of 856pp. 6½ × 9¼.
21873-2, 21874-0 Pa., Two-vol. set $17.00

WEBSTER'S AMERICAN MILITARY BIOGRAPHIES, edited by Robert McHenry. Over 1,000 figures who shaped 3 centuries of American military history. Detailed biographies of Nathan Hale, Douglas MacArthur, Mary Hallaren, others. Chronologies of engagements, more. Introduction. Addenda. 1,033 entries in alphabetical order. xi + 548pp. 6½ × 9¼. (Available in U.S. only)
24758-9 Pa. $11.95

LIFE IN ANCIENT EGYPT, Adolf Erman. Detailed older account, with much not in more recent books: domestic life, religion, magic, medicine, commerce, and whatever else needed for complete picture. Many illustrations. 597pp. 5⅜ × 8½.
22632-8 Pa. $8.50

HISTORIC COSTUME IN PICTURES, Braun & Schneider. Over 1,450 costumed figures shown, covering a wide variety of peoples: kings, emperors, nobles, priests, servants, soldiers, scholars, townsfolk, peasants, merchants, courtiers, cavaliers, and more. 256pp. 8⅜ × 11¼. 23150-X Pa. $7.95

THE NOTEBOOKS OF LEONARDO DA VINCI, edited by J. P. Richter. Extracts from manuscripts reveal great genius; on painting, sculpture, anatomy, sciences, geography, etc. Both Italian and English. 186 ms. pages reproduced, plus 500 additional drawings, including studies for *Last Supper, Sforza* monument, etc. 860pp. 7⅞ × 10¾. (Available in U.S. only) 22572-0, 22573-9 Pa., Two-vol. set $25.90

CATALOG OF DOVER BOOKS

THE ART NOUVEAU STYLE BOOK OF ALPHONSE MUCHA: All 72 Plates from "Documents Decoratifs" in Original Color, Alphonse Mucha. Rare copyright-free design portfolio by high priest of Art Nouveau. Jewelry, wallpaper, stained glass, furniture, figure studies, plant and animal motifs, etc. Only complete one-volume edition. 80pp. 9⅜ × 12¼. 24044-4 Pa. $8.95

ANIMALS: 1,419 COPYRIGHT-FREE ILLUSTRATIONS OF MAMMALS, BIRDS, FISH, INSECTS, ETC., edited by Jim Harter. Clear wood engravings present, in extremely lifelike poses, over 1,000 species of animals. One of the most extensive pictorial sourcebooks of its kind. Captions. Index. 284pp. 9 × 12. 23766-4 Pa. $9.95

OBELISTS FLY HIGH, C. Daly King. Masterpiece of American detective fiction, long out of print, involves murder on a 1935 transcontinental flight—"a very thrilling story"—NY Times. Unabridged and unaltered republication of the edition published by William Collins Sons & Co. Ltd., London, 1935. 288pp. 5⅜ × 8½. (Available in U.S. only) 25036-9 Pa. $4.95

VICTORIAN AND EDWARDIAN FASHION: A Photographic Survey, Alison Gernsheim. First fashion history completely illustrated by contemporary photographs. Full text plus 235 photos, 1840-1914, in which many celebrities appear. 240pp. 6½ × 9¼. 24205-6 Pa. $6.00

THE ART OF THE FRENCH ILLUSTRATED BOOK, 1700-1914, Gordon N. Ray. Over 630 superb book illustrations by Fragonard, Delacroix, Daumier, Doré, Grandville, Manet, Mucha, Steinlen, Toulouse-Lautrec and many others. Preface. Introduction. 633 halftones. Indices of artists, authors & titles, binders and provenances. Appendices. Bibliography. 608pp. 8⅜ × 11¼. 25086-5 Pa. $24.95

THE WONDERFUL WIZARD OF OZ, L. Frank Baum. Facsimile in full color of America's finest children's classic. 143 illustrations by W. W. Denslow. 267pp. 5⅜ × 8½. 20691-2 Pa. $5.95

FRONTIERS OF MODERN PHYSICS: New Perspectives on Cosmology, Relativity, Black Holes and Extraterrestrial Intelligence, Tony Rothman, et al. For the intelligent layman. Subjects include: cosmological models of the universe; black holes; the neutrino; the search for extraterrestrial intelligence. Introduction. 46 black-and-white illustrations. 192pp. 5⅜ × 8½. 24587-X Pa. $6.95

THE FRIENDLY STARS, Martha Evans Martin & Donald Howard Menzel. Classic text marshalls the stars together in an engaging, non-technical survey, presenting them as sources of beauty in night sky. 23 illustrations. Foreword. 2 star charts. Index. 147pp. 5⅜ × 8½. 21099-5 Pa. $3.50

FADS AND FALLACIES IN THE NAME OF SCIENCE, Martin Gardner. Fair, witty appraisal of cranks, quacks, and quackeries of science and pseudoscience: hollow earth, Velikovsky, orgone energy, Dianetics, flying saucers, Bridey Murphy, food and medical fads, etc. Revised, expanded In the Name of Science. "A very able and even-tempered presentation."—The New Yorker. 363pp. 5⅜ × 8. 20394-8 Pa. $6.50

ANCIENT EGYPT: ITS CULTURE AND HISTORY, J. E Manchip White. From pre-dynastics through Ptolemies: society, history, political structure, religion, daily life, literature, cultural heritage. 48 plates. 217pp. 5⅜ × 8½. 22548-8 Pa. $4.95

CATALOG OF DOVER BOOKS

SIR HARRY HOTSPUR OF HUMBLETHWAITE, Anthony Trollope. Incisive, unconventional psychological study of a conflict between a wealthy baronet, his idealistic daughter, and their scapegrace cousin. The 1870 novel in its first inexpensive edition in years. 250pp. 5⅜ × 8½. 24953-0 Pa. $5.95

LASERS AND HOLOGRAPHY, Winston E. Kock. Sound introduction to burgeoning field, expanded (1981) for second edition. Wave patterns, coherence, lasers, diffraction, zone plates, properties of holograms, recent advances. 84 illustrations. 160pp. 5⅜ × 8¼. (Except in United Kingdom) 24041-X Pa. $3.50

INTRODUCTION TO ARTIFICIAL INTELLIGENCE: SECOND, ENLARGED EDITION, Philip C. Jackson, Jr. Comprehensive survey of artificial intelligence—the study of how machines (computers) can be made to act intelligently. Includes introductory and advanced material. Extensive notes updating the main text. 132 black-and-white illustrations. 512pp. 5⅜ × 8½. 24864-X Pa. $8.95

HISTORY OF INDIAN AND INDONESIAN ART, Ananda K. Coomaraswamy. Over 400 illustrations illuminate classic study of Indian art from earliest Harappa finds to early 20th century. Provides philosophical, religious and social insights. 304pp. 6⅞ × 9⅞. 25005-9 Pa. $8.95

THE GOLEM, Gustav Meyrink. Most famous supernatural novel in modern European literature, set in Ghetto of Old Prague around 1890. Compelling story of mystical experiences, strange transformations, profound terror. 13 black-and-white illustrations. 224pp. 5⅜ × 8½. (Available in U.S. only) 25025-3 Pa. $5.95

ARMADALE, Wilkie Collins. Third great mystery novel by the author of *The Woman in White* and *The Moonstone*. Original magazine version with 40 illustrations. 597pp. 5⅜ × 8½. 23429-0 Pa. $9.95

PICTORIAL ENCYCLOPEDIA OF HISTORIC ARCHITECTURAL PLANS, DETAILS AND ELEMENTS: With 1,880 Line Drawings of Arches, Domes, Doorways, Facades, Gables, Windows, etc., John Theodore Haneman. Sourcebook of inspiration for architects, designers, others. Bibliography. Captions. 141pp. 9 × 12. 24605-1 Pa. $6.95

BENCHLEY LOST AND FOUND, Robert Benchley. Finest humor from early 30's, about pet peeves, child psychologists, post office and others. Mostly unavailable elsewhere. 73 illustrations by Peter Arno and others. 183pp. 5⅜ × 8½. 22410-4 Pa. $3.95

ERTÉ GRAPHICS, Erté. Collection of striking color graphics: *Seasons, Alphabet, Numerals, Aces* and *Precious Stones*. 50 plates, including 4 on covers. 48pp. 9⅜ × 12¼. 23580-7 Pa. $6.95

THE JOURNAL OF HENRY D. THOREAU, edited by Bradford Torrey, F. H. Allen. Complete reprinting of 14 volumes, 1837–61, over two million words; the sourcebooks for *Walden*, etc. Definitive. All original sketches, plus 75 photographs. 1,804pp. 8½ × 12¼. 20312-3, 20313-1 Cloth., Two-vol. set $80.00

CASTLES: THEIR CONSTRUCTION AND HISTORY, Sidney Toy. Traces castle development from ancient roots. Nearly 200 photographs and drawings illustrate moats, keeps, baileys, many other features. Caernarvon, Dover Castles, Hadrian's Wall, Tower of London, dozens more. 256pp. 5⅜ × 8¼. 24898-4 Pa. $5.95

CATALOG OF DOVER BOOKS

AMERICAN CLIPPER SHIPS: 1833-1858, Octavius T. Howe & Frederick C. Matthews. Fully-illustrated, encyclopedic review of 352 clipper ships from the period of America's greatest maritime supremacy. Introduction. 109 halftones. 5 black-and-white line illustrations. Index. Total of 928pp. 5⅜ × 8½.
25115-2, 25116-0 Pa., Two-vol. set $17.90

TOWARDS A NEW ARCHITECTURE, Le Corbusier. Pioneering manifesto by great architect, near legendary founder of "International School." Technical and aesthetic theories, views on industry, economics, relation of form to function, "mass-production spirit," much more. Profusely illustrated. Unabridged translation of 13th French edition. Introduction by Frederick Etchells. 320pp. 6⅛ × 9¼. (Available in U.S. only) 25023-7 Pa. $8.95

THE BOOK OF KELLS, edited by Blanche Cirker. Inexpensive collection of 32 full-color, full-page plates from the greatest illuminated manuscript of the Middle Ages, painstakingly reproduced from rare facsimile edition. Publisher's Note. Captions. 32pp. 9⅜ × 12¼.
24345-1 Pa. $4.95

BEST SCIENCE FICTION STORIES OF H. G. WELLS, H. G. Wells. Full novel *The Invisible Man*, plus 17 short stories: "The Crystal Egg," "Aepyornis Island," "The Strange Orchid," etc. 303pp. 5⅜ × 8½. (Available in U.S. only)
21531-8 Pa. $4.95

AMERICAN SAILING SHIPS: Their Plans and History, Charles G. Davis. Photos, construction details of schooners, frigates, clippers, other sailcraft of 18th to early 20th centuries—plus entertaining discourse on design, rigging, nautical lore, much more. 137 black-and-white illustrations. 240pp. 6⅛ × 9¼.
24658-2 Pa. $5.95

ENTERTAINING MATHEMATICAL PUZZLES, Martin Gardner. Selection of author's favorite conundrums involving arithmetic, money, speed, etc., with lively commentary. Complete solutions. 112pp. 5⅜ × 8½.
25211-6 Pa. $2.95

THE WILL TO BELIEVE, HUMAN IMMORTALITY, William James. Two books bound together. Effect of irrational on logical, and arguments for human immortality. 402pp. 5⅜ × 8½.
20291-7 Pa. $7.50

THE HAUNTED MONASTERY and THE CHINESE MAZE MURDERS, Robert Van Gulik. 2 full novels by Van Gulik continue adventures of Judge Dee and his companions. An evil Taoist monastery, seemingly supernatural events; overgrown topiary maze that hides strange crimes. Set in 7th-century China. 27 illustrations. 328pp. 5⅜ × 8½.
23502-5 Pa. $5.95

CELEBRATED CASES OF JUDGE DEE (DEE GOONG AN), translated by Robert Van Gulik. Authentic 18th-century Chinese detective novel; Dee and associates solve three interlocked cases. Led to Van Gulik's own stories with same characters. Extensive introduction. 9 illustrations. 237pp. 5⅜ × 8½.
23337-5 Pa. $4.95

Prices subject to change without notice.
Available at your book dealer or write for free catalog to Dept. GI, Dover Publications, Inc., 31 East 2nd St., Mineola, N.Y. 11501. Dover publishes more than 175 books each year on science, elementary and advanced mathematics, biology, music, art, literary history, social sciences and other areas.